CORE JAVA

(First Edition)

By

Er. Nagesh Jaitak

B.Tech in CSE

PTU,JALANDHAR

CONTENTS

Unit 1: Introduction to Java

Objectives

After studying this unit, you will be able to:

- Describe the features of Java
- Explain the Java programming techniques
- List the keywords used in Java
- Discuss constants in Java
- Describe variables in Java

- Explain data types in Java
- Discuss operators in Java
- Describe expressions in Java

Introduction

Sun Microsystems began the development of Java confidentially in 1991, which was later released to the public in 1995. Java is comparatively a new and exciting technology. Interestingly, it was developed with the aim of providing solutions for household appliances such as telephones. From there on, it has developed into a fully functional programming language. Java can be integrated directly onto a Web page as an applet (a Java program that can be included in an HTML page). This makes the Internet a much more dynamic and interesting place to gather information, do business, or just have fun! In fact, this dynamic aspect of Java is what initially sparked interest in many researchers and engineers and therefore, resulted in its popularity.

Did you Know?

Sun Microsystems originally wanted to name Java as "OAK". But it could not do so as that name was already taken by Oak Technologies. Other names that were suggested were "Silk and "DNA". Ultimately, the name "Java" was selected because it gave the Web a "jolt", and Sun intended to abstain from names that sounded very technical.

Java is an important object-oriented programming language that is used in the software industry today. Object Oriented Programming is also known as OOP. Objects are the basic elements of object-oriented programming. OOPS (Object Oriented Programming System) is used to describe a computer application that comprises multiple objects connected to each other. It is a type of programming language in which the programmers not only define the data type of a data structure (files, arrays and so on), but also define the behavior of the data structure. Therefore, data structure becomes an object, which includes both data and functions.

Just like any other programming language, Java programs consist of some basic elements such as keywords, constants, variables, data types, operators and expressions that help a programmer to create logical programs, and some important features such as platform-independence which enables it to run on any computer platform.

1.1 Features of Java

Let us understand Java better by understanding its important features. Following features of Java make it an important programming language:

1. *Platform Independent*: The write-once-run-anywhere approach towards programming is one of the key features of Java that makes it a powerful programming language. The programs written on one platform can run on any platform, irrespective of the hardware. But the hardware platform used to execute Java programs must have the Java Virtual Machine (JVM).

2. *Simple*: There are various features that make Java a simple language, which can be easily learnt and effectively used. Java does not use pointers explicitly, thereby making it easy to write and debug programs. Java is capable of delivering a bug free system due to its strong memory management. It also has an automatic memory allocation and de-allocation system in place.

3. *Object Oriented*: To qualify as an object-oriented language, a language must exhibit four characteristics:

    (a) *Inheritance*: It is the technique of creating new classes by making use of the behavior of the existing classes. This is done by extending the behavior of the existing classes just by adding additional features as required, thus bringing in reusability of existing code.

    (b) *Encapsulation*: It refers to the bundling of data along with the methods that act on that data.

(c) *Polymorphism:* Polymorphism, which means one name multiple forms, is the ability of a reference variable to change behavior according to the instance of the object that it holds.

(d) *Dynamic binding:* It is the method of providing maximum functionality to a program by resolving the type of object at runtime.

Although the forerunners of Java, like Objective C and C++, fulfill the above four characteristics, they are not completely object-oriented, because they follow structured programming as well as object-oriented programming. However, Java is completely object-oriented since everything in Java is an object.

4. *Robust:* Java supports some features such as automatic garbage collection, strong memory allocation, powerful exception handling, and type checking mechanism. The compiler checks the program for errors and the interpreter checks for any run time errors, thus preventing the system crash. These features make Java robust.

5. *Distributed:* The protocols like HTTP and FTP, which are extensively used over the Internet are developed using Java. Programmers who work on the Internet can call functions with the help of these protocols and can secure access to the files that are present on any remote machine on the Internet. This is made possible by writing codes on their local system itself.

6. *Portable:* The feature 'write-once run anywhere, anytime' makes Java portable, provided that the system has JVM. Java standardizes the data size, irrespective of the operating system or the processor. These features make Java a portable language.

7. *Dynamic:* A Java program also includes significant amount of runtime information that is used to verify and resolve access to objects at runtime. This allows the code to link dynamically in a secure and appropriate manner.

8. *Secure:* Memory pointers are not explicitly used in Java. All programs in Java are run under Java execution environment. Therefore, while downloading an applet program using the Internet, Java does not allow any virus or other harmful code to access the system as it confines it to the Java execution environment.

9. *Performance:* In Java, a program is compiled into an intermediate representation, which is called Java bytecode. This code can be executed on any system that has a JVM running on it. Earlier attempts to achieve cross-platform operability accomplished it at the cost of performance. Java bytecode is designed in such a manner that it is easy to directly translate the bytecode into the native machine code by using a just-in-time compiler. This helps in achieving high performance.

10. *Multithreaded:* The primary objective that led to the development of Java was to meet the real-world requirement of creating interactive and networked programs. In order to accomplish this, Java provides multithreaded programming, which permits a programmer to write programs that can do many things simultaneously.

11. *Interpreted:* Java programs can be directly run from the source code. The source code is read by the interpreter program and translated into computations. The source code generated is platform independent. As an interpreted language, Java has an error debugging facility that can trace any error occurring in the program.

12. *Architecture Neutral:* Java is also known as an architectural neutral language. In this era of networks, easy migration of applications to different computer systems having different hardware architectures and/or operating systems is necessary. The Java compiler generates an object file format that is architecture neutral. This permits a Java application to execute anywhere on the network and on many different processors, given the presence of the Java runtime system.

These features of Java have made it a popular programming language.

1.2 Programming in Java

Any Java program starts with the class name and every class definition starts with an opening curly brace ({) and ends with the corresponding curly brace (}). After the opening brace, the main line of the program begins. For example:

public static void main (String args[ ])

This line defines the main method of the program. Then, the output line is added to the program.

System.out.println ("Hello");

This statement is used to print the output on the screen.

Example: Program to print "Programming".

```
Class ExampleProgram
{
    public static void main (String args[ ])
    {
        System.out.println("I am a simple program");
    }
}
```

Output:
I am a simple program.

In this example,

(a) A class **ExampleProgram** is created using the keyword **class**.

(b) In the class **ExampleProgram**, the **main( )** method of the class is called, using the **public static void main (String args[ ]) statement**.

(c) Then, in the **main( )** method, **System.out.println("I am a simple program");** statement is used to print the output "I am a simple program" on the screen.

1.2.1 Structure of a Java Program

A Java program is divided into six sections, as given in the table 1.1.

Table 1.1: Structure of a Java Program

Comment Lines' Section	(Suggested Section)
Package Section	(Optional Section)
Import Statements' Section	(Optional Section)
Interface Statements' Section	(Optional Section)
Class Definitions' Section	(Optional Section)
Main Method Class Section	(Essential Section)

Let us understand the six sections of Java program in detail:

1. *Comment Lines' Section*: This section consists of a collection of comment lines, which includes the name, author and other details of the program.

2. *Package Section*: This section consists of the first statement that is permitted in a Java file. This first statement is called as the package statement. It gives the name of the package and provides information to the compiler that all the classes defined in the respective program are related to this package.

 Example: package employee;

   In this statement, **package** is the keyword and **employee** is the package name.

3. *Import Statements' Section:* This section consists of import statements, which can be used for accessing the classes that are related to other packages.

 Example: package employee.salary;

   In this statement, the interpreter is instructed to load the **salary** class that is contained in the package **employee**.

4. *Interface Statements' Section:* This section consists of interface statements that are used only if the multiple inheritance feature is to be implemented in the program.

5. *Class Definitions' Section:* This section consists of the definition of the classes used in a program.

6. *Main Method Class Section:* This section consists of the definition of the main method in the program. The main method enables the creation of objects of different classes and the communication between these objects.

After understanding the basic elements, that is, keywords, variables, constants, and data types, a programmer can easily implement this structure while developing programs in Java.

1.2.2 Compiling, Interpreting and Running the Program

It is necessary to convert the program into a format that can be understood by the JVM, before any computer with a JVM can interpret and run the program. The process of compiling a Java program involves obtaining the programmer-readable text in the program file (source code) and translating it to bytecode. Bytecode are nothing but the platform-independent instructions for the JVM.

After compiling and interpreting, a program is run. To run a program, follow the instructions given below:

1. Go to command prompt.

2. Select the drive (and the folder) where the program file is saved.

3. Compile, Interpret, and Run the program.

To further understand these instructions, refer to the following example.

 Example: Type the following program and save it in a file with the name "FirstProgram.java"

```
class FirstProgram
{
    public static void main(String args[ ])
    {
```

```
        System.out.println("My first program in Java");
    }
}
```

In this example,

    (a) A class **FirstProgram** is created using the **class** keyword.

    (b) In the class **FirstProgram,** the **main( )** method of the class is called.

    (c) In this **main( )** method, **System.out.println("My first program in Java");** statement is used to print the output **"My first program in Java"** on the screen.

The process to compile and run the above-given example is described in the following steps

1.   In the command prompt, type the directory name, and the file name.

    Example: c:\jdk\programs>javac FirstProgram.java

        Here, **c:** is the drive, wherein **jdk** directory is stored. The **'programs'** is the directory, where the program which is to be run is saved with the name **FirstProgram.java**.

2.   After this step is followed by the programmer, a file called as **FirstProgram.class** is created by the compiler in the directory **programs** (check that it is there!). This class created by the compiler has the program but in bytecode form, which is ready to run. Run the program as given in the example below.

    Example: C:\jdk\programs > java FirstProgram

        The output of this program is:

        My first program in Java

These steps, if followed properly, will give the desired output to the user.

Compilation

At the command line on UNIX and DOS shell operating systems, the Java compiler for the example provided in section 1.2 is invoked as follows:

        javac ExampleProgram.java

Here, javac is the command and ExampleProgram.java is the name of the class (or program file).

    In Java, the class name and the file name must be the same. Also, since Java is case sensitive, one should always provide same case letters for both file name and class name.

Once the program is successfully compiled into Java bytecode, it is ready to be interpreted and run on any JVM. Interpreting and running of a Java program involves invoking the JVM bytecode interpreter. It transforms the Java bytecode to platform-dependent machine codes so the computer can understand and run the program.

Interpretation

At the command line on UNIX and DOS shell operating systems, the Java interpreter is invoked as follows:

java ExampleProgram

Running

At the command line, you should see the output that is given within the System.out.println statement in the program.

I am a simple program

Task Write a simple Java program to print "Java".

1.3 Keywords

Keywords are predefined identifiers set aside by Java for a particular purpose. These keywords cannot be used as names for variables, classes, methods or as identifiers or tokens. All keywords must be written in lowercase letters. At present, there are fifty defined keywords in the Java language. Table 1.2 gives a list of the Java keywords.

Table 1.2: Keywords in Java				
abstract	Continue	assert	default	Boolean
do	Double	byte	else	case
enum	Catch	extends	char	final
class	Finally	const	float	for
goto	If	implements	import	instanceof
int	Interface	long	native	new
package	Private	protected	public	return
short	Static	strictfp	super	switch
synchronized	This	throw	throws	transient
try	Void	volatile	break	while

The keywords in Java have a predefined meaning and they perform distinct functions. Some of the keywords and their functions are listed below:

1. *abstract*: This Java keyword is used for declaring a method without providing that method's implementation, in a method declaration.

2. *assert*: This Java keyword is used for making a condition's assumed value explicit.

3. *boolean*: This Java keyword is used for relating to an expression or variable, which can have only either a true or false value.

4. *break:* This Java keyword is used to restart the execution of a program at the statement that is just after the current enclosing block or statement. In case this keyword has a label just after it, the program restarts execution at the statement just after the enclosing labeled statement or block.

5. *byte:* This Java keyword is an 8-bit integer. It is used for declaring the value that a method will return, an expression, or variable of type **byte**.

6. *case:* This Java keyword is used for defining a set of statements. This keyword is generally used with the switch statement.

7. *catch:* This Java keyword is used for handling the exceptions occurring in a program above **try** keyword.

8. *char:* This Java keyword is used for declaring an expression, value that a method will return, or variable of type character.

9. *class:* This Java keyword is used for defining the implementation of an object of a particular kind.

10. *continue:* This Java keyword is used for continuing the program at the end of the body of the current loop.

11. *default:* This Java keyword is used for defining a set of statements to begin the execution. This is to be used in case no suitable match is found for the value that is defined by the enclosing switch statement, among the values indicated by a case keyword in the switch statement.

12. *do:* This Java keyword is used for declaring a loop that will repeat a block of statements. The exit condition of this loop is indicated with the while keyword. The loop's execution will happen once, before the evaluation of the exit condition.

13. *double:* This Java keyword is a 64-bit floating point value. This keyword is used for declaring an expression, the value a method will return, or variable that is of type double-precision floating point.

14. *else:* This Java keyword is used for testing the condition. If the test condition indicated by the **if** keyword evaluates to false, a statement or block of statements are executed. The **else** keyword is used for defining such statement or block of statements.

15. *enum:* This Java keyword is used for declaring an enumerated type.

16. *extends:* This Java keyword is used for specifying the super class in a class declaration, and also for specifying one or more super interfaces in the declaration of an interface.

17. *final:* This Java keyword is used for defining an entity once, which cannot be changed or derived later. Also, no class can be inherited from a **final** class, and all methods defined in a **final** class are completely final.

To know the functions of other keywords, refer http://en.wikipedia.org/wiki/List_of_Java_keywords

or http://www.cafeaulait.org/course/week2/09.html

Caution   Java supports some reserved words such as true, false, and null. As these are reserved words, not keywords, they cannot be used as names for programs.

1.4 Constants

Unlike keywords that are predefined and kept aside by Java for special purpose, constants are values that do not change during the program execution. The keyword **final** is used to declare the constants.

Two types of constants are provided by Java, as shown in the figure 1.1.

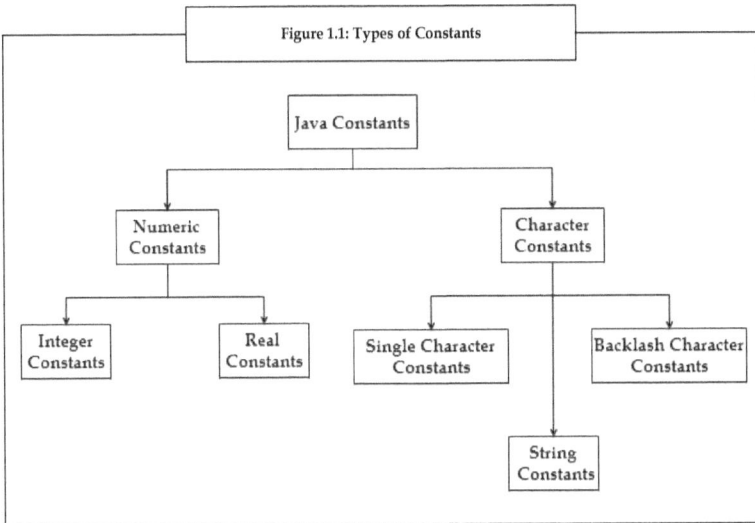

Figure 1.1: Types of Constants

Numeric Constants

Numeric constants refers to just the numbers used in a program. There are two types of numeric constants:

1. **Integer Constants**: Integer numeric constants consist of digits 0 through 9, which may be written with or without a + or – sign. These constants do not contain decimal values. Integer constants can be classified into three types:

    (a) *Decimal Integer Constant:* This type of integer constant comprises a collection of digits, 0 through 9, which may or may not have a minus sign before them. For example, 132, -312, 0, 54321. Spaces in between the numbers, commas, and non-digit characters are not allowed between digits.

     Example: 14 250, 10,000, $500- are unacceptable numbers in Java.

    (b) *Octal Integer Constant:* This type of integer constant comprises a combination of digits, from the digits 0 through 7. Every such combination must be preceded by a 0.

     Example: 032, 0, 0534, 0451 are octal integer constants.

    (c) *Hexadecimal Integer Constant:* This type of integer constant has a 0x or 0X before a set of digits. This set may also comprise alphabets, A through F, or a through f, which represents the numbers 10 through 15.

     Example: 0X3, 0x8f, 0x are hexadecimal integer constants.

    Octal and hexadecimal numbers are rarely used in programming languages.

2. ***Real Constants***: Real numeric constants are the constants that are used for representing the quantities that differ continuously such as temperatures, heights, distances, prices, and so on. Such constants contain fractional parts.

 Example: 0.082, -0.65, 42.31 are real numeric constants.

It is also possible that there are no digits before the decimal point or no digits after the decimal points.

Character Constants

Character constants refer to those constants that contain either single character or in the form of a string of characters. We can classify character constants into the following three types:

1. ***Single Character Constants***: Single character constant, also called as character constants, consist of a single character that is enclosed within a pair of single quotation marks.

 Example: '6', ' ', 'B' are single character constants.

2. ***String Constants***: String constants refer to those character constants where a series of characters are enclosed within a pair of double quotation marks. These characters may be alphabets, digits, blank spaces, or special characters.

 Example: "Hello World", "1991", "2+1", "X" are string constants.

3. ***Backslash Character Constants***: Backslash constants are those character constants that are used in the output methods. Table 1.3 shows a list of backslash character constants.

Table 1.3: Backslash Character Constants

Backslash Character Constant	Meaning
'\n'	New line
'\\'	Backslash
'\f'	Form feed
'\b'	Back space
'\r'	Carriage return
'\t'	Horizontal tab
'\''	Single quote
'\"'	Double quote

In table 1.3, every backslash character constant consists of two characters, but represents only one character. Such character combinations are also referred to as escape sequences.

1.5 Variables

Just like constants, variables also play an important role in programming languages. In Java, a variable is the basic storage unit, which stores the states of objects. A variable has a name and data type, where the data type specifies the type of value that the variable can hold.

1.5.1 Declaring and Initializing Variables

Before using a variable in a program, it must be declared first. Declaration involves specifying the data type of the variable. Initializing is the process of assigning value to the variable. The value must be compatible with the variable's data type.

NAGESH JAITAK

The declaration of a variable is as follows:

<data type> <identifier> [= value] [, identifier[= value]….];

Here, values enclosed in < > are required values, while values enclosed in [ ] are optional.

A variable name may comprise digits, alphabets, dollar characters, and underscore (_), provided the following conditions are satisfied:

1. Do not begin a variable name with a digit.

2. Use uppercase and lowercase appropriately, as they are different. For example, *Sum* is not similar to *sum* or *SUM*.

3. Do not use keywords as a variable name.

4. Do not provide white spacing in between the variable name.

1.5.2 Primitive Variables vs Reference Variables

There are two types of variables in Java, namely, reference variables and primitive variables. They differ in meaning and in function.

Primitive variables are variables with primitive data types. They store data in the actual memory location where the variable is, whereas reference variables are variables that store the address in the memory location. They point to the memory location where the data is actually present.

While comparing primitive variables, the actual values are compared, but while comparing reference variables, the addresses are compared.

When a variable of a certain class is being declared, a reference variable to the object with that particular class is actually being declared.

 Example: Suppose there are two variables with data type **int** and **String**.

int num = 10;

String name = "Hello";

Let the memory location of the variable **num** be 1001 and the memory location of variable **name** be 1563. But, the data at the memory location 1563 points to another address where the data of the variable is actually stored.

For the primitive variable **num**, the data is on the actual location of where the variable is. For the reference variable **name**, the variable just holds the address of the location where the actual data is present.

1.6 Data Types

In any programming language, whether it is Java or C++ or any other programming language, a data type is said to be a collection of data having values with predefined characteristics. For example, integer, floating point number, and character string are data types.

In Java, every variable has a data type associated with it. Data types indicate the type and the size of values that can be stored in that variable.

In Java, data types can be categorized into two types, as given in figure 1.2.

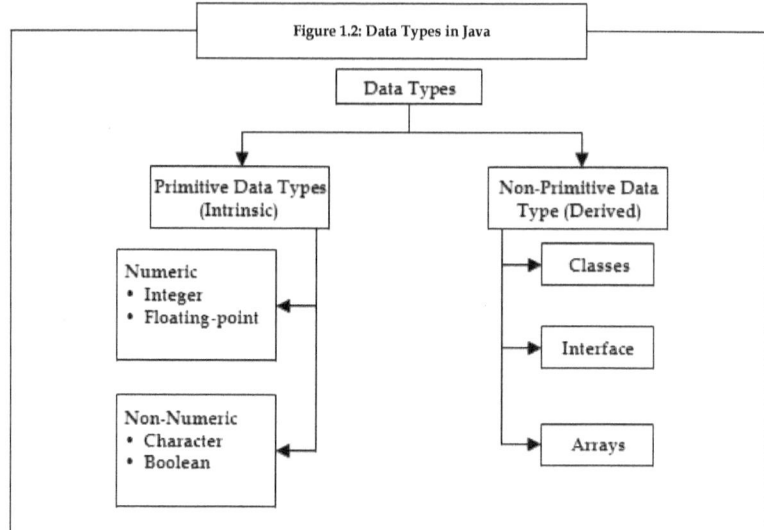

Figure 1.2: Data Types in Java

Primitive Data Types (Built-in Data Types)

Generally, few data types come in-built in a programming language. The range of values for a given data type is described by the language itself. The way in which the values are processed and stored by the computer is also specified by the language. Primitive data types are used to categorically store information or data that may or may not be interchangeable, and that is built-in. They are predefined by Java and are named by a reserved keyword. Each class is also considered to have its own data type (a class also acts as a data type).

We can classify primitive data types into two types:

1. Numeric data types

2. Non-numeric data types

Numeric Data Types

Data types that contain numbers only, are called as numeric data types. These data types are of two types:

1. *Integer Data Types*: Data types that hold numbers, such as, 122, -92, and 4639, are called as integer data types. In Java, there are four types of integers, namely, byte, short, int, and long.

Table 1.4 shows the size and range of these integer types.

Name or Type	Range	Integer Length
byte	-128 to 127	8 bits
short	-32,768 to 32,767	16 bits
int	-2,147,483,648 to 2,147,483,647	32 bits
long	-9,223,372,036,854,775,808 to 9,223,372,036,854,775,807	64 bits

Table 1.4: Integer Types and their Ranges

The ranges given in the table 1.4 for different integral types help a programmer to limit the size of the variable and use the appropriate data type for the variables.

2. **Floating Point Data Types**: Data types that are used to hold the fractional parts are called as floating point data types. There are two types of floating point data types, namely, **float**, and **double**. Floating point types have double as default data type. A floating-point literal has either a decimal point or one of the following:

    (a) F or f -- float

    (b) D or d – double

    (c) E or e -- add exponential value

 Example: 8.14 -- A simple floating-point value (a double)

5.02E24 -- A large floating-point value

6.718F -- A simple float size value

523.4E+306D -- A large double value with redundant D

In these examples, the 24 after the E in the second example is implicitly positive. That example is equivalent to 5.02E+24.

Table 1.5 shows the size and range for the floating-point data types.

Name or Type	Range	Float Length
Float	1.4e-045 to 3.4e+038	32 bits
Double	-4.9e-324 to 1.8e+308	64 bits

Table 1.5: Floating Point Types and their Ranges

The ranges given in the table 1.5 for different floating point types help a programmer to limit the size of the variable and use the appropriate data type for the variables.

Non-numeric Data Types

Data types that are not numeric in nature are called as non-numeric data types. There are two types of non-numeric data types:

1. *Character Data Type*: Character data type is called as char, which is used for storing the character constants in the memory.

2. *Boolean Data Type*: Boolean data type is used whenever any particular condition is to be tested at the time of program execution. Boolean data type can take only two values, either true or false. This data type is denoted by the keyword Boolean.

Non-primitive Data Types

Non-primitive data types are those data types that are not in-built and are user–defined. These data types are also called as derived data types. We can classify these data types into the following three types:

Classes

A **class** is a user-defined data type that defines the basic components of a Java program called **objects**. The creation of these objects is done by the classes and communication between these objects is done using the **methods** (objects use methods).

Interface

An **interface** is basically a type of **class**, which consists of methods and variables, but defines only the **abstract** methods and the **final** fields. As there is no concept of multiple inheritance in Java, interfaces are used to support the concept of multiple inheritance in Java.

Arrays

An **array** is a collection of data items that are related or contiguous and share a common name, which is followed by a number called as index number.

 Example: In **salary[10]**, salary is the array name, and 10 is the index number.

1.6.1 Enum Types

An enum type or an enumerated type refers to a type whose fields comprise a fixed set of constants.

 Example: Days of week, Planets in solar system contains fixed set of constants.

The following example shows how to declare an enum type.

 Example: enum Week

```
{
    // enum constants
    MONDAY, TUESDAY, WEDNESDAY, THURSDAY, FRIDAY, SATURDAY,
    SUNDAY
}
```

The keyword enum indicates that the declaration is in an enumeration type. The fields of enum need to be given in capital letters.

1.7 Operators

In real time, programs are required to perform a lot more than just simple input and output operations. All computer languages provide tools to perform some predefined operations. To facilitate this, there are different types of operators provided in Java. These are arithmetic operators, relational operators, logical operators and conditional operators. These operators are assigned a certain kind of precedence, so that the compiler evaluates the operations on the basis of their precedence. This helps when there are multiple operators in one statement.

1.7.1 Arithmetic Operators

Like the mathematical expressions used in algebra, Java also allows the use of arithmetic operators to be used in mathematical expressions.

Table 1.6 shows the basic arithmetic operators that can be used while writing Java programs.

Table 1.6: Arithmetic Operators and their Functions

Operator	Use	Description
+	op1 + op2	Adds op1 to op2
*	op1 * op2	Multiplies op1 to op2
/	op1 / op2	Divides op1 by op2
%	op1 % op2	Computes the remainder after division of op1 by op2
-	op1 - op2	Subtracts op2 from op1

The operations performed by the arithmetic operators shown in the table 1.6 are the same as that of the mathematical operators.

1.7.2 Increment and Decrement Operators

In addition to the basic arithmetic operators, Java also includes a unary increment operator (++) and unary decrement operator (--). The function of increment and decrement operators is to increase and decrease a value stored in a number variable by 1 respectively.

For example, the expression **count = count + 1;** increments the value of count by 1. This is equivalent to, **count++**.

The increment and decrement operators can precede the operand or follow the operand as shown in table 1.7.

Table 1.7: Increment and Decrement Operators		
Operator	Use	Description
++	op++	Increments op by 1; evaluates to the value of op before it is incremented
++	++op	Increments op by 1; evaluates to the value of op after it is incremented
--	op--	Decrements op by 1; evaluates to the value of op before it is decremented
--	--op	Decrements op by 1; evaluates to the value of op after it is decremented

When used before the operand, it increments or decrements the variable by 1, and then the new value is used in the expression in which it appears.

Example: int i = 15,

int j = 4;

int k = 0;

k = ++j + i; // will result to k = 5+15 = 20

When the increment and decrement operators are used after the operand, the old value of the variable will be used in the expression, where it appears first. The next subsequent appearance of the variable will have the incremented value.

Example: int i = 15,

int j = 4;

int k = 0;

k = j++ + i; // will result in k = 4+15 = 19

1.7.3 Relational Operators

Relational operators compare two or more values and determine the relationship between those values. These operators actually determine equality and ordering. The result of evaluation is the **Boolean** value **true** or **false**.

Table 1.8 shows the list of relational operators used in Java.

Table 1.8: Relational Operators		
Operator	Use	Description
>	op1 > op2	op1 is greater-than op2
>=	op1 >= op2	op1 is greater-than or equal to op2
<	op1 < op2	op1 is less than op2
<=	op1 <= op2	op1 is less than or equal to op2
==	op1 == op2	op1 is equal to op2
!=	op1 != op2	op1 is not equal to op2

The operations performed by the relational operators such as less-than, equal-to, greater-than and so on are shown in table 1.8. They are the same as the mathematical relational operators.

1.7.4 Logical Operators

Java also has logical operators, which act on one or two Boolean operands yielding a Boolean result. There are six logical operators, **&&** (short-circuit AND), **&** (logical AND), **||** (short-circuit OR), **|** (logical inclusive OR), **^** (logical exclusive OR), and **!** (logical NOT).

The basic syntax for a logical operation is:

a <operator> b

In the above syntax, **a** and **b** can be Boolean expressions, variables or constants and <operator> is either &&, &, ||, | or ^ operator.

&& (short-circuit AND) and & (logical AND)

Table 1.9 gives the truth table for logical operators && and & for all the possible combinations of **a** and **b**. A truth table is a dissection of a logical function which lists all possible values that the function can attain. A truth table consists of many rows and columns where the top row represents the logical variables and combinations, leading up to the final result.

Table 1.9: Truth Table for & and &&		
a	b	Result
TRUE	TRUE	TRUE
TRUE	FALSE	FALSE
FALSE	TRUE	FALSE
FALSE	FALSE	FALSE

Table 1.9 gives the logic identity of the logic values (**a** and **b** here).

There is a main difference between **&&** and **&** operators, which states that **&&** assists short-circuit evaluations (or partial evaluations), while **&** does not.

Example: Given an expression:

test1 && test2

In this expression, **&&** will evaluate the expression test1 and immediately return a false value, if **test1** is false. This is because, the operator never evaluates **test2**, as the result of the operator will be false, regardless of the value of **test2**. In contrast, & operator always evaluates both **test1** and **test2** before returning an answer.

| | (short-circuit OR) and | (logical inclusive OR)

Table 1.10 gives the truth table for logical operators | | and | for all the possible combinations of a and b.

	Table 1.10: Truth Table for	and			
a	**b**	**Result**
TRUE	TRUE	TRUE
TRUE	FALSE	TRUE
FALSE	TRUE	TRUE
FALSE	FALSE	FALSE

Table 1.10 gives the logical values for the logical variables **a** and **b**, by performing 'or' logical operation.

The basic difference between | | and | operators is that | | supports short-circuit evaluations, while | does not.

 Example: Given an expression:

test1 | | test2

In this expression, | | will evaluate the expression **test1** and immediately return a true value if **test1** is true. If **test1** is true, the operator never evaluates **test2**, because the result of the operator will be true, regardless of the value of **test2**. In contrast, the | operator always evaluates both **test1** and **test2** before returning an answer.

^ (logical exclusive OR)

Table 1.11 gives the truth table for ^ for all the possible combinations of **a** and **b**.

Table 1.11: Truth Table for ^		
a	b	Result
TRUE	TRUE	FALSE
TRUE	FALSE	TRUE
FALSE	TRUE	TRUE
FALSE	FALSE	FALSE

Table 1.11 gives the logical values for the logical variables **a** and **b** by performing 'XOR' logical operation. In 'XOR' logical operation, if one operand is true and the other is false, then only the final result of an exclusive OR (XOR) operation is true. Note that both operands must always be evaluated in order to determine the result of an exclusive OR.

! (logical NOT)

The logical NOT, another logical operator, takes in one argument. This argument can be an expression, variable or constant. Table 1.12 gives the truth table for !.

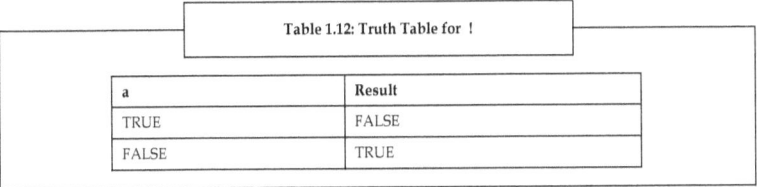

Table 1.12: Truth Table for !	
a	Result
TRUE	FALSE
FALSE	TRUE

Table 1.12 gives the logical value for the variable **a** by performing logical 'NOT' operation. This operator is used by the programmer to reverse the value of a **Boolean** expression or **Boolean** condition.

 Example: If **a** is false, !**b** is true.

1.7.5 Conditional Operator '?:'

The conditional operator ?: is a ternary (three-way) operator. It takes three arguments that together form a conditional expression. The syntax for an expression using a conditional operator is:

expression1? expression2: expression3

In the above syntax, **expression1** is a Boolean expression whose result must either be true or false. If **expression1** is true, **expression2** is the value returned. If it is false, then **expression3** is returned.

1.7.6 Operator Precedence

Java operator precedence determines the precedence of operators, that is, which operator has to be evaluated first. Just like operator precedence in algebra, Java operator precedence defines the order in which various operators are evaluated.

Operator precedence gives the compiler's order of evaluation of operators so as to achieve unambiguous result. Sensible use of parenthesis (operator precedence) can make the programs easy to read, even if the expressions are complicated. Use of parentheses also raises the precedence of the operator of the operations that are inside those parentheses. This is often necessary to obtain the desired result.

Table 1.13 shows the precedence of operators, where first row shows the highest precedence and last row the lowest precedence.

Table 1.13: Operator Precedence

Highest			
( )	[ ]	.	
++	--	~	!
*	/	%	
+	-		
>>	>>>	<<	
>	>=	<	<=
==	!=		
&			
^			
|			
&&			
||			
?:			
=	op=		
Lowest			

 Example: Consider a complicated expression:

6%2*5+4/2+88-10

It can be re-written by placing some parenthesis based on operator precedence, as:

((6%2)*5)+(4/2)+88-10

Operators are used for the manipulation of primitive data types. In Java, we can categorize operators as unary (taking one argument), binary (taking two arguments), or ternary (taking three arguments). In Java, the use of operators enables a programmer to evaluate mathematical functions. The operator precedence defined in Java also helps to clearly express mathematical functions and expressions.

1.8 Expressions

Like operators, expressions are also an important element of programming. An expression is said to be a piece of program code, which is used for representing or computing a value. An expression can be in the form of a literal, a variable, a function call, or many of these things integrated together with the help of operators such as + and >. The user can assign the expression value to a variable, use it as a parameter in a subroutine call, or combine it with other values into an expression that is more complex.

In Java, expression has an important type, that is, arithmetic expression. When variables, constants, and operators are combined, according to the syntax of the programming language (here, Java), an arithmetic expression is formed.

Notes

Java does not support any operator for exponentiation.

The user can evaluate expressions with the help of an assignment statement of the following syntax:

variable = expression;

In this syntax, **variable** states any valid name of a variable in Java. While evaluating any statement, the evaluation of the expression is done first and then the previous value of the variable that is on the left-hand side (in the syntax) is replaced by the result of this evaluation. Before the evaluation of the expression, ensure that all the variables that are used in the expression are given values. A sample of evaluation statement is:

x = a/b*c;

In this sample, the variables **x**, **a**, **b**, and **c** that are used must be defined before their usage in the expression.

If an arithmetic expression is without parentheses, its evaluation will start from left and move towards right using the rules of the operators' precedence. In Java, arithmetic operators have two different priority levels:

1.   High priority: * / %

2.   Low priority: + -

The basic evaluation is done in two steps:

1.   In the first step, the operators that are of high priority (if any) are applied whenever encountered.

2.   In the second step, the operators that are of low priority (if any) are applied whenever encountered.

Expressions Having Type Conversion

In any Java expression, the user can mix two or more distinct types of data as long as they are congruent with each other. During evaluation of the expression, rules of type conversion must be adhered to. If two different operands are used in an expression, automatic conversion of the **lower** type to the **higher** type takes place before the evaluation.

 Example: If three variables of type **byte, short**, and **int**, are used in an expression, the result is always converted to **int**.

Table 1.14 shows the reference chart for type conversion.

	char	byte	short	int	long	float	double
Table 1.14: Automatic Type Conversion Chart							
char	int	Int	int	int	long	float	double
byte	int	Int	int	int	long	float	double
short	int	Int	int	int	long	float	double
int	int	Int	int	int	long	float	double
long	long	long	long	long	long	float	double
float	float	float	float	float	float	float	double
double	double	double	double	double	double	double	double

After the evaluation of expression, first the conversion of the result to the variable type, which exists on the left-side of the operator sign is done, then the value is assigned to it.

Value Casting

Apart from automatic type conversions, there is one concept related to type conversion, which is casting a value (value casting). Sometimes, a type conversion needs to be forced in a way that is different from the automatic conversion.

 Example: Calculation of the ratio of girls to boys in a city, where the number of girls and boys are declared as integers.

ratio = girls_number/boys_number

As the number of girls and boys are already defined as integers, the fractional part or the decimal part of the calculated result, or ratio, will not be visible. Thus, correct figure will not be presented in ratio. User can solve this problem by converting locally any one of the two given variables to float as given below:

ratio = (float) girls_number/boys_number

The **girls_number** is converted to floating point by using the **(float)** operator for evaluating the expression. Thereafter, the automatic conversion rule is used to perform the division operation in floating point mode. Thus, the fractional part is also retained in the result. The (float) operator does not affect the value of the variable **girls_number** and the type int of **girls_number** will remain same in the other parts of the program.

This process of local conversion of a value is called as value casting.

Case Study

Success Story of University of Alabama Birmingham Medical Center (UABMC)

Many medical institutions depend on film-based medical images for diagnosis. One such nstitution is the University of Alabama Birmingham Medical Center (UABMC). The process that is used (film-based images) is completely manual, but inefficient and costly.

This system is inefficient and costly because of the printing costs, proper film storage costs, time that is lost in waiting for film delivery, and re-imaging patients due to lost films. To enhance the patient care and remove these unessential costs, UABMC made a decision. The decision was to become a "film optional" medical center using digital medical images. Dr. Bart Guthrie, associate professor of Neurosurgery at UABMC said, "The computer-assisted surgery needs made it necessary that we evolve the capability for electronic image capture and distribution. Afterwards, we felt that the delivery of care for any specialty is improved by the immediate access to medical images, not just Neurosurgery. Now, an electronic Clinical Image Management System (CIMS) is evolved, which makes the care provider able to make decisions that are active and informed and is altering the way we work, by opening up new channels of answer and care. We enhance patient care and reduce costs by shifting to the digital care." We enhance patient care and reduce costs by shifting to the digital care."

To attain the medical images to the care-point, UABMC and ComFrame, a consulting company that specializes in Java technology consulting and healthcare products, created an application based on the Java platform. This application permits physicians and clinicians to see medical images on any desktop anywhere in the medical center. By using the Java technology, this application can fetch DICOM images from a collection that is based on the name of the patient, medical record number of the patient, or type and date range, and shifts those images to a desktop Java software-based application that permits the clinicians to see and maneuver the images. The enhanced imaging and performance capabilities of the Java platform were important to build an effective application and to assist physicians and clinicians to offer faster and better patient care.

"We used Java technology for this application for meeting some needs and specifications," says Dr. Guthrie. "For becoming effective, we had to meet the requirements of clients using MACs, Windows, and various distinct UNIX platforms. The application users demand high with respect to performance requirements that we are working hard to meet on all platforms."

Dr. Gary York of ComFrame agreed and said "Java is a wonderful object-oriented platform; it permitted us to develop a strong, scalable answer instantly with a small team and the enhancements of the Java platform really gave energy to this application. For example, we required the enhanced image maneuver capabilities of the Java API (Application Programming Interface – Interface used to access an application or service from a program). As medical images are large and complicated and with JDK 1.1.6 (Java Development Kit – Software development environment of Java), some images maneuver operations on the client took seconds. With the Java platform, we have decreased the maneuver time to sub-seconds - essentially real time with the new fast indexed color models. We also use Java threads on the client for making the anticipated performance higher, as a result of which the visual feedback is immediate. This upgrade of the performance is acknowledged by the physicians who rely on this application for quick decision making. Many Java components were used to write the application, so we value that these Java components are a part of Java platform's core APIs."

This electronic image management process is now used by over 120 physicians and clinicians, and UABMC finally plans to offer access for up to 1000 users. The UABMC will achieve this target, in part by evolving a Web-based applet version of the application to target users such as general practitioners, who use imaging as an integrated part of their work only occasionally.

Questions

1. What is the purpose of the application developed by UABMC and ComFrame?

2. Why was only Java technology used to create this application?

1.9   Summary

- Java was developed by Sun Microsystems initially to offer solutions for household appliances. But, finally it evolved as a fully functional programming language.

- Java has many features such as platform-independence, simplicity, object-oriented capability, portability, and so on that differentiates it from other programming languages and makes it important.

- Structure of a Java program consists of comment lines' section, package section, import statements' section, interface statements' section, class definitions' section, and main method class section.

- All Java keywords have predefined meanings and distinct functions. They must be written in lowercase letters.

- Constants refer to the values that do not change during the program execution.

- Constants are of two types, namely, numeric constants (integer constants, real constants) and character constants (single character constants, string constants and backslash character constants).

- A variable is the basic storage unit used for storing the states of objects, which has a name and data type.

- Every variable has a data type associated with it, which indicate the type and the size of values that can be stored in that variable.

- Data types are of two types – primitive and non-primitive data types. Primitive data types are in-built in any programming language and are used for categorically storing information or data that may or may not be interchangeable. Non-primitive data types are defined by the user.

- In Java, operators such as arithmetic operators, relational operators, logical operators, and conditional operators are used to perform some predefined operations.

- An expression is said to be a piece of program code used for representing or computing a value, which can be in the form of a literal, a variable, a function call, or many of these things integrated together with the help of operators such as + and >.

- As per the rule of type conversion, in case two distinct operands are used in an expression, automatic conversion of the 'lower' type to the 'higher' type takes place before the evaluation.

1.10   Keywords

FTP: File Transfer Protocol

HTTP: Hyper Text Transfer Protocol. A networking protocol for collaborating, distributed, hypermedia information systems developed by the Internet Engineering Task Force and the World Wide Web consortium.

OOP: Object Oriented Programming

Operators' Precedence: Order in which the operators are executed during expression evaluation

1.11   Self Assessment

1.   State whether the following statements are true or false:

    (a) OOP is a type of programming language in which the programmers define the data type of a data structure and the behavior of the data structure.

    (b) OOPS is used to describe a computer application that does not comprise multiple objects that are connected to each other.

    (c) Inheritance is the technique of creating new classes by making use of the behavior of the existing classes.

(d) The feature 'write-once run anywhere, anytime' makes Java portable, without any other requirement on any system.

(e) The **break** keyword is used to restart the execution of the program at the statement that is just after the current enclosing block or statement.

(f) String constants are those character constants, which are used in the output methods.

(g) An array is a collection of data items that are related or contiguous and share a common name, which is followed by a number called as index number.

(h) Relational operators actually determine equality and ordering.

(i) Decision control structures are Java statements that allow a programmer to select and execute specific blocks of code while bypassing other sections.

(j) The constructor's name and name of the class in which it is created are same.

2. Fill in the blanks:

(a) _____is the mechanism that enables us to combine the information and provide abstraction.

(b) All keywords must be written in _____letters.

(c) The keyword _____is used to declare the constants.

(d) While comparing primitive variables, the actual values are compared, but while comparing reference variables, the _____are compared.

(e) _____data types are used to categorically store information or data that may or may not be interchangeable, and that is built-in.

(f) When variables, constants, and operators that are arranged according to the syntax of the programming language are combined, an _____ is formed.

1.12 Review Questions

1. Analyze different features of Java, which has made Java an important programming language.

2. "Structure of a Java program consists of six sections." Justify.

3. "Keywords are predefined identifiers set aside by Java for a particular purpose, which cannot be used as names for variables, classes or methods." Discuss.

4. "Constant refers to the values that do not change throughout the program." Discuss with the help of its types.

5. "Declaration of a variable involves specifying the data type of the variable." Discuss.

6. "Primitive variables and reference variables are different from each other." Justify.

7. "In every Java program, different types of data are used." Elaborate.

8. "All computer languages provide tools to perform some predefined operations." What are these tools and what are their types?

9. "Value casting is an important concept used in Java". Elaborate.

10. "An interface is a type of class." Comment.

11. "Increment and decrement operators are used to increment or decrement values." Elaborate.

12. "To run a program in Java, the user can use command prompt." Discuss.

Answers: Self Assessment

1. (a) True   (b) False   (c) True   (d) False   (e) True

    (f) False   (g) True   (h) True   (i) True   (j) True

2. (a) Encapsulation   (b) Lowercase   (c) Final

    (d) Addresses   (e) Primitive   (f) Arithmetic Expression

1.13 Further Readings

Books

Balagurusamy E. Programming with Java_A Primer 3e. New Delhi

Schildt. H. Java 2 The Complete Reference, 5th ed. New York: McGraw-Hill/Osborne.

Online link

www.roseindia.net/java/java-introduction/java-features.shtml

www.tech-faq.com/java-data-types.html - United States

http://java.sun.com/j2se/press/bama.html

Unit 2: Fundamentals of OOP

Objectives

After studying this unit, you will be able to:

- Define classes, objects and methods

- Explain class inheritance

- Describe constructors

Introduction

OOP (Object Oriented Programming) is a popular programming style that is chosen by almost all the software companies. OOP, also commonly referred as OOPs, is a modern method of organizing and developing programs. It is not necessary that every language supports the OOP concepts. Some of the programming languages that support OOP concepts are C++, Smalltalk, Object Pascal, and Java, which is purely object oriented. In OOP, data is a crucial element in the process of program development and is not permitted to be used freely in the system. Unlike procedural programming, in the OOP programming model, programs are organized around objects and data rather than actions and logic.

Java is an object oriented programming language and therefore to understand the functionality of OOP in Java, we first need to understand several fundamentals related to objects. These include class, method, objects and so on.

2.1 Introduction to Classes, Objects and Methods

OOP is based on classes and objects. A **class** is said to be a collection of objects with common behavior (set of actions that can be performed by an object). In OOP, a problem is divided into a number of entities, which are called as **objects**.

 Example: Suppose a program is written for calculating the number of accounts in a bank. In this program, **bank** can be a class, and **customer** and **account** can be two objects of that class.

After this division, data and methods are created around these methods, as given in the figure 2.1.

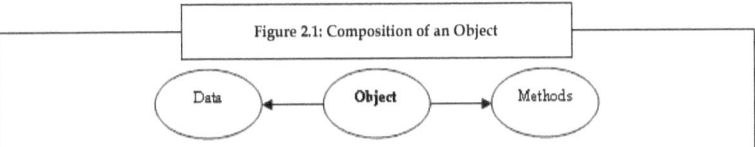

Figure 2.1: Composition of an Object

A method is declared within the body of a class, and it operates on data. An object is equal to the combination of data and methods. An object's data is accessible only by the methods that are related to that object.

2.1.1 Defining a Class

In object oriented programming, there is a basic structure named as a **class**, which is a blueprint or a template from which objects are created. A **class** is a framework, which represents a general object. An instance of a class represents an individual **object**. The **object** defines the behaviors and properties of the **class**. A class contains properties and methods.

Structure of a Class

Classes contain the following three parts:

1. *Class Definition:* A class is defined using the keyword **class**, which is followed by the name of the class.

2. *Instance and Class Variables' Definition:* After defining the class, variables to be used in the program are defined (within the curly braces).

3. *Methods' Definition:* After defining the variables, methods (if any) to be used in the program are defined.

General Syntax of a Class

class Classname

    {

        Variables' definition;

        Methods' definition;

    }

Example: Program to illustrate the method used to define a class.

public class MyPoint

{

public int a = 2;

public int b = 4;

void displayPoint( )

{

  System.out.println ("Printing the point");

  System.out.println (a + " " + b);

}

}

In this example,

1. First, a class **MyPoint** is created and is declared as public.

2. In this class,

    (a) Two integer variables **a** and **b** are declared, and assigned the values **2** and **4** respectively. The **public** keyword that is associated with these variables indicates that any other class can access these variables.

    (b) Then, the **displayPoint ( )** method of the class is called to display the point.

    (c) In this **displayPoint ( )** method,

        (i) The System.out.println("Printing the point"); statement is used to print **Printing the point** on the screen.

        (ii) Then, the values of **a** and **b** are printed on the screen.

2.1.2 Creating Objects

Objects are the basic elements of object oriented programming. There are many objects in the real world around us such as cars, buses and so on. The objects are categorized based on their properties and behaviors.

 Example: The object car has a set of properties like manufacturer and color. Its behavior includes the speed and mileage.

General Syntax of an Object

Classname c1 = new Classname( );

In this syntax, **Classname** is the name of the class, **c1** is the name of the new object and **new** is the operator used to create the new object.

In the example class MyPoint, there is no main method because it is not a complete application. To use the MyPoint class, another class has to be created. A MyPointDemo class creates the object of the class MyPoint and calls its method.

Example: Program to illustrate the method to define an object.

```
public class MyPointDemo
{
public static void main(String args[ ])
{
MyPointDemo p1 = new MyPointDemo( );
p1.displayPoint( );
}
}
```

In this example,

1. First, a class **MyPointDemo** is created.
2. In this class,

    (a) The **main( )** method of the class is called.

    (b) In this **main( )** method,

        (i) A new object p1 of the class MyPointDemo is created by using the new operator.

        (ii) The displayPoint ( ) method of the MyPoint class is called to display the object p1.

    Finally, the new object **p1** is printed on the screen.

2.1.3 Object Class

All Java classes are derived from **object class**, which is the base class for all the classes used in Java, either directly or indirectly. Some important methods that are used in the object class are:

1. *equals:* The **equals** method is used for the comparison between two objects, that is, to check for equality. If these objects are equal, this method will return **true**, otherwise it will return **false**.

2. *getClass:* The **getClass** method is used for querying the object of the class for different information such as, name of the class, its super class, etc.

3. *toString:* The **toString** method is used to display the string representation of an object, which completely relies on that object.

 Example: The **String** representation of a **float** object is the float value displayed as text.

2.1.4 Introducing Nested and Inner Classes

A **nested class** is a class that is defined within another class. Nested classes enhance the readability and maintainability of a code.

An example of nested class is as follows:

 Example: Program to show a nested class.

```
class MainClass  // this is the main class
{
    int a= 50;  // variable of type int is declared and is assigned a value of 50
    void display( ) // a method display of type void is declared
    {
// A new instance of NestedClass is created
        NestedClass  nest = new NestedClass( );
/* the object of the nested class calls the method present in the NestedClass class
*/
        nest.showData( );
    }


    class NestedClass
    {
        void showData( )  // a method display of type void is declared
        {
            System.out.println("The value stored in variable a of main class is:" +a);
        }
    }
}
class SampleNestedClass
{
    public static void main(String args[ ])
    {
// A new instance of MainClass is created
        MainClass mc = new MainClass( );
/* the object of MainClass calls the display method present in the class MainClass
*/
        mc.display( );
    }
}
```

Output:

The value stored in variable a of main class is: 50

In this example,

The MainClass is the outer class and the NestedClass is the inner class. The inner class accesses the instance variable of the outer class and displays the value that it holds. In the main( ) method, the mc object of the MainClass is created to invoke the method display( ). The display( ) method invokes the showData( ) method present in the NestedClass class and outputs 50.

Nested classes are of two types as shown in the figure 2.2.

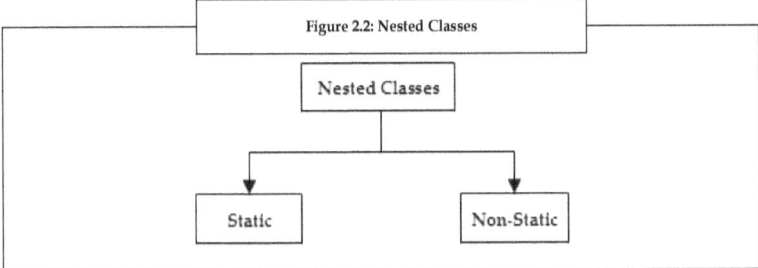

Figure 2.2: Nested Classes

Nested Classes

Static   Non-Static

1. *Static Nested Classes:* Static nested classes are those nested classes, which are declared static using the keyword **static**. Such classes can access the instance variables indirectly, with the help of an object of the encapsulating class.

2. *Non-Static Nested Classes:* Non-static nested classes, also called as **inner classes**, are those nested classes that are declared without using the **static** keyword. Such classes can directly access the instance variables.

Nested classes and inner classes are commonly used in Java programs.

2.1.5 Wrapper Class

A wrapper class is a class that envelops the value of every primitive data type. This class represents primitive data types in their equivalent class instances.

 Example: An **integer** data type can be represented as an **Integer Class** instance.

 Notes   The user cannot create any subclass of a wrapper class, as this class is always declared final.

Features of Wrapper Classes

Wrapper classes have the following features:

1. Methods defined within wrapper classes are **static.**

2. Wrapper classes are immutable, that is, we cannot change any value after assigning it to a wrapper class.

Table 2.1 shows the list of primitive data types and their equivalent wrapper classes.

Table 2.1: Primitive Data Types and their Equivalent Wrapper Classes	
Primitive Data Type	**Wrapper Class**
Boolean	java.lang.Boolean
byte	java.lang.Byte
char	java.lang.Character
double	java.lang.Double
float	java.lang.Float
int	java.lang.Integer
long	java.lang.Long
short	java.lang.Short

In table 2.1, wrapper classes include the name of the wrapper library and the wrapper name.

 Example: In **java.lang.Integer**, **java.lang** is the name of the wrapper library, and **Integer** is the wrapper name.

2.1.6 Abstract Class

An abstract class refers to a class that is declared by using the abstract keyword. It helps to organize classes based on common methods. It helps to put the common method names in one abstract class without having to write the actual implementation code.

Abstract class cannot be instantiated. However, they can be extended into sub classes. These sub classes usually provide implementations for all of the abstract methods included in the abstract class. The abstract classes may or may not comprise abstract methods.

 Example: public abstract class GrapObj

{

    abstract void draw(); //declaration of abstract method

}

2.1.7 Final Class

A class can be declared as a final class, by using the **final** keyword, which means that the class cannot have a subclass.

 Example: final class Test

Here, **final** is the keyword that is used to declare the **class Test** as a final class.

In a final class, all the methods are completely declared as **final**, but it is not necessary that all the data types are also **final.**

 Do not try to create subclasses of final classes, as it will cause an error and will not be permitted by the compiler.

2.1.8 Using Super Class

A super class is defined as a class that can be extended into subclasses. The concept of super classes is used in inheritance, where a new class (a subclass) is created by deriving some existing behavior and states of another class (super class), using the **extends** keyword. This method is usually used to enhance code reusability.

 Example: public **class** dog **extends** animal

Here, dog is the **subclass** of **super class** animal.

2.1.9 Adding Variables and Methods to a Class

In a program, first a class is defined and then the variables to be used in that program are defined within that class. In a Java program, the **static** keyword is used to specify that the variable is a class variable. This indicates that this variable has only one copy and it is associated with a particular class. Variables that are not declared static are called as instance variables.

Did you know?   Variables must be initialized before being used in a program.

In any program, methods are used for the manipulation of data used in that program. For adding a method to any class, specify that method within that class' declaration. A method can be declared with or without parameters.

 Example: Program to find out the square root of a number.

```
import java.lang.Math
class Squroot
{
    public static void main (String args[ ])
    {
        double a = 4;
        double b;
        b = Math.sqrt (a);
        System.out.println ("Square root is: " +b);
    }
}
```

Output:

Square root is: 2.0

In this example,

1. First, the **java.lang.Math** package is imported, which consists of methods to perform mathematical operations.

2. Then, a class **Squroot** is created.

3. In this class **Squroot,**

    (a) The main( ) method of the class is called.

    (b) In the main( ) method,

        (i) Two variables a and b, of type double are declared. Also, variable a is assigned a value of 4.

        (ii) A method Math.sqrt ( ); is called for calculating the square root of value assigned to variable a. The result is then assigned to b.

        (iii) Finally, System.out.println ("Square root is: " +b); statement is used to print the value of b on screen.

2.2 Constructors

Constructor is a special type of method, by which an object can initialize itself on its creation. The constructor's name and name of the class in which it is created are same. No return type (not even void) is given by a constructor because the instance of the class itself is returned by the constructor. Generally, a constructor is used for providing initial values to the instance variables that are defined by the class, or for carrying out procedures (if any) needed for creating a fully formed object. In Java, all classes consist of constructors, whether they are defined or not. This is because a default constructor is automatically provided for the initialization of all the member variables to zero. However, after the definition of a new constructor, the default constructor is not used.

Some constructors require one or more parameters to be added to them in the same way that they are added to a method. Such constructors are called as parameterized constructors. To add parameters to a constructor, declare the parameters inside the parentheses after the name of the constructor.

 Example: Program to illustrate the use of a simple constructor in a program.

```
class DemoClass
{
int a;
DemoClass( )
{
  a = 20;
}
}
class DemoConstr
{
  public static void main(String args[ ])
  {
    DemoClass c1 = new DemoClass( );
    DemoClass c2 = new DemoClass( );
    System.out.println(c1.a + " " + c2.a);
  }
}
```

Output:

20 20

In this example,

1. First, a class **DemoClass** is created.

2. In the **DemoClass**,

    (a) An integer **a** is declared.

    (b) Then, the constructor for DemoClass, that is, DemoClass( ) is called. In this constructor DemoClass( ), the instance variable **a** of DemoClass is assigned a value of 20.

3. A new class **DemoConstr** is created.

4. In the class **DemoConstr,**

    (a) The main( ) method of the class is called.

    (b) In the main( ) method,

        (i) Two new objects c1 and c2 of DemoClass are created by calling the constructor DemoClass( ), and by using the new keyword.

        (ii) Finally, the value of a is called for the object c1, and is printed on the screen.

2.3 Class Inheritance

Inheritance is one of the object-oriented concepts. It is a process, where one object inherits the properties of another. Similarly, class inheritance means that a class derives a set of properties and methods of a parent class or base class.

To inherit a base class to its subclass, a keyword **extends** is used in the subclass definition.

Java provides two types of inheritance. They are:

1. Simple Inheritance

2. Multi-level Inheritance

2.3.1 Simple Inheritance

When a subclass is derived directly from its parent class or super class, it is known as simple inheritance. In simple inheritance, there is only a subclass and its super class. It is also referred to as single inheritance or one-level inheritance.

 Example: Program to illustrate the use of simple inheritance in a program.

```
public class Super class
{
  public void A( )
  {
    System.out.println("Print the super class method");
  }
}
```

```
class Subclass extends Super class
{
    public static void main(String args[ ])
    {
        Subclass s1 = new Subclass( );
        s1.A( );   }
}
```

Output:

Print the super class method

In this example,

1. First, a parent class or base class **Super class** is created.

2. In the **Super class, A( ) method** is called.

3. In the A( ) method, System.out.println("Print the super class method"); statement is used to print the super class method statement on the screen.

This application cannot be run, as it does not have main method. So, further additions are made to the program:

1. A class **Subclass** is the inherited from the class **Super class** by using the **extends** keyword.

2. In the class **Subclass,**

    (a) The main( ) method of the class is called.

    (b) In the main( ) method,

    (i) The Subclass s1 = new Subclass( ); statement is used to create a new object s1 of class Subclass using the new operator.

    (ii) Then, s1.A( ); statement is used to call the method of the Super class is called with the Subclass object s1.

Finally, the program prints the method of the super class.

Task

Write a simple inheritance program to show one subclass **manager** and its super class **employee.**

2.3.2 Multi-level Inheritance

Multi-level inheritance was introduced to enhance the concept of inheritance. When a subclass is derived from another subclass or derived class, it is known as the multi-level inheritance. In multi-level inheritance, the subclass is the child class for its super class and this super class is the subclass for another super class. Multi-level inheritance can go up to any number of levels.

Example: Program to illustrate the use of multi-level inheritance in a program.

```
class P
{
    int e;
    int f;
    int get(int a, int b)
    {
        e=a; f=b;
        return(0);
    }
    void Show( )
    {
        System.out.println(e);
    }
}
class Q extends P
{
    void Showq( )
    {
        System.out.println("Q");
    }
}
class R extends Q
{
    void display( )
    {
        System.out.println("R");
    }
    public static void main(String args[ ])
    {
        P p = new P( );
        p.get(7,10);
        p.Show( );
    }
}
```

Output:

7

In this example,

1. First, a class **P** is created, wherein integers **e** and **f** are declared.

2. In the class **P**,

    (a) The get( ) method is called to get the values of integers **a** and **b**.

    (b) In the get( ) method,

    (i) The values of **a** and **b** are assigned to **e** and **f** respectively.

    (ii) Then, return(0); statement is used to return the value of 0.

    (iii) The **Show( )** method is then called, wherein **e** is printed on the screen.

3. A new class **Q** is created as a subclass of the class **P** using the **extends** keyword.

4. In the class **Q**, void **Showq( )** method is called, wherein **Q** is printed on the screen.

5. A new class **R** is created as a subclass of the class **Q** using the **extends** keyword.

6. In the class **R**,

    (a) The display( ) method is called, wherein R is printed on the screen.

    (b) Then, the main( ) method of the class is called.

    (c) In the main( ) method,

    (i) A new object p of the class P is created using the new keyword.

    (ii) Then, the int get( ) method of the class P is called with the values for the variables in the parameter list, that is, 7 and 10.

    (iii) At last, Show( ) method is called to show the value of p.

    (iv) Finally, the output 7 is displayed on the screen.

Java does not support multiple inheritance. That is, we cannot inherit from more than one class. This is because multiple inheritance has many drawbacks.

Let us consider an example.

Example: class X

```
{
    int p = 10;
}
class Y extends X
{
    int p;
}
class Z extends X
{
```

```
      int p = 30;
}
// Now in case java support multiple inheritance (which it does not)
class D extends Y, Z
{
    public static void main(String [ ] args)
    {
        D q = new D( );
        System.out.println(q.p);
    }
}
```

In this example,

If two classes are inherited, which p will be printed? The value of p in class Z or the value of p in class Y. This is the problem with multiple inheritance. This is known as a **diamond problem**. This is why Java does not support multiple inheritance.

The above example is only one main drawback of multiple inheritance.

The main reason for omitting multiple inheritance from the Java language is to keep the language as simple as possible. The creators of Java wanted a language that most developers could learn without extensive training. Therefore, they worked to make the language as similar to C++ as possible without carrying over C++'s unnecessary complexity. The designers were of the opinion that multiple inheritance causes more problems and confusion than it solves.

However, at times we may want to derive from two or more classes. Therefore, the designers of Java chose to allow multiple interface inheritance through the use of interfaces. We shall learn more about Interfaces in Unit no 6.

Lab Exercise   Write a program to find out the square root of 16.

2.4 Summary

- OOP is a famous programming style consisting of object-oriented concepts such as classes and objects.
- The structure of a class comprises three sections, namely, class definition section, instance and class variables' definition section, and methods' definition section.
- Object class is the main class in any Java program.
- When a class consists of class or classes created within that class, it is called as a nested class.
- A wrapper class encases the values of all the primitive data types, by creating a wrapper for every type.
- A final class is a class with no subclasses and declared with the keyword **final.**
- A super class is the main class from which subclasses are derived or inherited.
- Variables and methods are added to a class right after the definition of a class in a program.

- Class inheritance is defined as a process in which, a class derives a set of properties and methods of a parent class or base class. To inherit a base class to its subclass, a keyword **extends** is used in the subclass definition. It can either be simple or multi-level.

- Constructor is a method that is used to construct a new object in a class, whose name is same as of the class it is created in. If the constructor has parameters too, then it is called as a parameterized constructor.

2.5 Keywords

Initialize: Allocate an initial value to a java program.

Instance Variable: A variable that is relevant to a single instance (an object belonging to a class is an instance of that class) of a class.

Primitive Data Types: Special group of data types that represents a single value such as numbers, characters, or Boolean values.

Wrapper Library: Library of wrappers of all the primitive data types.

2.6 Self Assessment

1. State whether the following statements are true or false.

    (a) Data is not a crucial element in the process of program development.

    (b) An object is a collection of classes with common behavior.

    (c) In a program, after the class definition, variables to be used in the program are defined.

    (d) Object class is the foundation class for all the classes.

    (e) Classes created within other classes are called as nested classes.

    (f) The user can create subclasses of a wrapper class.

2. Fill in the blanks:

    (a) The _____keyword is used to declare a class as a final class.

    (b) Class inheritance means that a class derives a set of _____and methods of a parent class or base class.

    (c) For adding _____to a constructor, they are declared inside the parentheses after the name of the constructor.

3. Select a suitable choice for every question.

    (a) Which keyword is used to create a subclass of a super class?

        (i) implements

        (ii) extends

        (iii) new

        (iv) getClass

    (b) The primitive data type Boolean comes under the _____ wrapper class.

        (i) java.lang.Boolean

        (ii) java.lang.Byte

        (iii) java.lang.Character

        (iv) java.lang.Double

(c) Which keyword is used to create a new object of any class?

   (i) equals

   (ii) new

   (iii) getClass

   (iv) toString

2.7 Review Questions

1. "In OOP, a problem is divided into a number of entities". Discuss.

2. "Structure of a class consists of three parts". Justify with an example.

3. "All Java classes are derived from object class, which is the base class for all the classes used in Java, either directly or indirectly". Discuss the methods associated with the object class.

4. "Nested classes enhance the readability and maintainability of code". Comment.

5. "A wrapper class is a class that envelops the value of every primitive data type." Discuss.

6. "A super class is defined as a class that can be extended into subclasses" Justify with an example.

7. "In any program, methods are used for the manipulation of data used in that program". Discuss.

8. "To inherit a base class to its subclass, a keyword **extends** is used in the subclass definition". Discuss.

Answers: Self Assessment

1. (a) False   (b) False   (c) True   (d) True   (e) True   (f) False

2. (a) final   (b) Properties   (c) Parameters

3. (a) extends   (b) java.lang.Boolean   (c) new

2.8 Further Readings

Books

Balagurusamy E. Programming with Java 3e Primer. New Delhi: Tata McGraw Publishers.

Online link

http://docstore.mik.ua/orelly/java/javanut/ch03_01.htm

http://www.roseindia.net/help/java/s/super-class.shtml

http://www.java-tips.org/java-se-tips/java.lang/what-is-a-final-class.html

Unit 3: Control Structures

Objectives

After studying this unit, you will be able to:

- Explain the use of decision control structures

- Illustrate the repetition control structures

- Describe the different branching statements

Introduction

The process of program development requires thorough understanding of the control structures that are used for the proper execution of programs. We know that the execution of statements in a program flows from top to bottom, in the order of their appearance in the program. Control structures however, break up the execution flow by implementing decision making, looping, and branching, making your program to conditionally execute particular code-blocks.

OOP concepts and control structures when combined together lead to effective and executable programs.

A program can manipulate its execution sequence and make choices during the execution of the program code. In Java, these choices are made with the help of execution control statements. The execution control statements allow a programmer to change the order in which the statements in the program are executed.

A program is said to be a group of statements, which is executed in the same sequential order in which they appear. Control statements, control structures, or control constructs refer to the statements that control the execution of these statements. These control structures in Java are categorized into three parts, that is, decision control structures, repetition control structures and branching statements.

These three control structures comprise various statements and loops that control the execution of the statements in a program.

3.1 Decision Control Structures

Decision control structures are Java statements that allow a programmer to select and execute specific blocks of code while bypassing other sections. These structures are also called as conditional statements.

All conditional statements evaluate to true or false for a conditional expression to determine their execution path.

Example: Consider an example of an expression, a == b. In this example, the conditional operator == is used to find out if the value of **a** is equivalent to the value of **b**. The expression returns either **true** or **false**. Any relational operator can be used in a conditional statement.

 Notes Java does not allow a number/integer to be used as a Boolean unlike C and C++ (where a non-zero value is assigned to 'true' and zero is assigned to 'false'). In order to use a non-Boolean variable in a Boolean condition such as if (a), the variable must first be converted into a Boolean value by using a conditional expression such as if(a != 0).

Decision control structures can be classified into four parts as given in the figure 3.1.

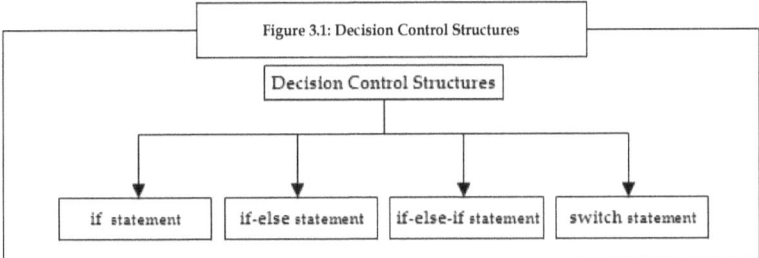

Figure 3.1: Decision Control Structures

'if' Statement

In Java, **if** statement is one of the decision control constructs. The **if** statement indicates that a statement or a set of statements will be executed, if and only if a certain Boolean expression/condition is true.

Syntax of 'if' Statement:

    if (condition/Boolean expression)

    statement1;

    or

    if(condition/Boolean expression)

    {

        statement1;

        statement2;

        . . .

    }

As given in the syntax, each statement may be a single statement or a compound statement. The condition/Boolean expression is an expression that returns a Boolean value. The **if** statement signifies that the execution of a statement (or a block of code) depends on the evaluation result of the Boolean expression, that is, a statement (or a block of code) will be executed, if and only if a certain Boolean statement is true.

Example: Program to illustrate the use of **if** statement in Java.

```
class If1
{
  public static void main (String args[ ])
  {
    int grade = 68;
    if( grade > 60 )
    {
System.out.println("Grade is greater than 60!");
    }
    System.out.println("Grade is less than 60!");
  }
}
```

Output:

Grade is greater than 60!

In this example,

1. First, a class If1 is created.
2. In this class,
    (a) The main( ) method of the class is called.
    (b) In this main( ) method,
        (i) An integer constant grade is declared, and is assigned a value of 68.
        (ii) Then, an if condition (grade>60) is evaluated.
        (iii) If grade is greater than 60, then the statement Grade is greater than 60! is executed, otherwise the statement outside the if block "Grade is less than 60!" is executed.

    Here, the statement **Grade is greater than 60!** is executed, as the value of grade is 68 (> 60).

Notes   The condition part of a statement must evaluate to a Boolean value. It implies that the execution of the condition should either result to true or false.

'if-else' Statement

The if-else statement is used, when the execution depends on the evaluation result of a condition or Boolean expression. It means that if the condition is true, a certain statement is executed, but if the condition is false a different statement is executed.

Syntax of 'if-else' Statement:

if (condition)

  statement 1;

else

  statement 2;

or

if (condition)

{

  statement 1;

  statement 2;

  . . .

}

else

{

  Statement 1;

  Statement 2;

  . . .

}

As per the above syntax, a condition is checked, if it is true, then the statements within the **if** block are executed, else the statements within the **else** block are executed.

 Example: Program to illustrate the use of **if-else** statement in Java.

```
class Test1
{
public static void main (String args[ ])
{
 int grade = 68;
 if( grade > 60 )
 {
   System.out.println("First Division");
 }
 else
 {
```

```
System.out.println("Not first division");
      }
    }
  }
```

Output:

First division

In this example,

1. First, a class Test1 is created.

2. In this class,

    (a) The main( ) method of the class is called.

    (b) In this main( ) method,

        (i)   An integer constant grade is declared, and is assigned a value of
              68.

        (ii) Then, an **if** condition (grade > 60) is evaluated.

        (iii) If grade is greater than 60, then the statement in the **if** block
              First division is executed, otherwise the statement in the else
              block Not first division is executed.

        Here, the statement **First division** is executed, as the value of grade is 68
        (> 60).

Task Write a program to accept the marks of a student and then print the result of a student, that is, **pass** or **fail**.

'if-else-if' Statement

The statement that is present in the **if** or **else** clause of an **if-else** statement, can be another **if** or **if-else** statement. This process is called nesting. An **if** or **if-else** statement that includes another **if** or **if-else** statement is called a **nested if** statement. This cascading of statements enables a programmer to make more complex selections. It becomes necessary to use more than one **if-else** statement in nested form for writing a program having a series of decisions or conditions.

Nested if-else statements are very common in programming. An important thing to remember while nesting if-else statements is that an else statement always belongs to the nearest if statement that is within the same block as the else and that which is not already associated with an else.

Syntax of if-else-if Statement:

```
if (condition1)
{
    if (condition2)
            {
                statement1;
            }
            else
            {
                statement2;
            }
    }
    else
    {
        if (condition3)
        {
            statement3;
        }
        else
        {
            statement4;
        }
    }
}
```

As per the above syntax, in **nested if-else** statement, **condition1** is the first condition to be evaluated. If it evaluates to true, then **condition2** is evaluated. If **condition2** is true, then **statement1** is executed, but, if it is not true, **statement2** is executed. If **condition1** is false, then **condition3** is evaluated. If **condition3** is true, then **statement3** is executed, but if it is not true, **statement4** is executed.

The **if** statement that come within another **if** statement is called an **inner if** statement, and an **if** statement that contains another **if** statement is called an **outer if** statement. Thus, in the above syntax, the **if** statement that tests **condition1** is an **outer if** statement, and the **if** statement that tests **condition2** and **condition3** is an **inner if** statement.

Notes

There can be many else-if blocks after an if-statement. The else-block is optional and can be excluded.

Example: Program to illustrate the use of **if-else** statement in Java.

```
class Grade
{
    public static void main (String args[ ])
    {
    int grade = 68;
    if( grade > 90 )
            {
                    System.out.println("Excellent");
            }
            else
            {
    if( grade > 60 )
    {
      System.out.println("Very good");
    }
    else
    {
      System.out.println("Sorry you failed");
    }
            }
        }
    }
```

Output:

Excellent

In this example,

1. First, a class **Grade** is created.
2. In this class,
   (a) The **main( )** method of the class is called.
   (b) In this main( ) method,
       (i) An integer constant grade is declared, and is assigned a value of **68.**
       (ii) Then, an **if** condition **(grade > 90)** is evaluated.

(iii) If the condition is **true,** then the statement in the first **if** block
Excellent is executed, otherwise the condition **grade > 60** in the else
block is evaluated.

(iv) If the **grade > 60** condition is **true,** statement in the **if** block
Very good is executed, otherwise the statement in the **else**
block **Sorry you failed** is executed.

Here, the statement within the first **if** block **"Excellent"** is executed.

Errors Made While Using the "if-else" Statement

While writing programs, programmers may commit a few common mistakes while using **if-else** statements. Few errors that occur while using **if-else** statements are:

1. The condition inside the if statement does not directly evaluate to a Boolean value. Therefore, we must give only an expression that evaluates to a Boolean value.

Example:

```
//CORRECT
int number=0;
if(number >= 0)
{
//some statements here
}
//INCORRECT
int number = 0;
if( number )
{
//some statements here
}
```

In this example, the variable number in the above example does not hold a Boolean value. Thus, it is incorrect to use a number like this, that is, no condition is being checked in the if block.

2. Sometimes, programmers use = instead of == for comparing two variables, values, and so on.

Example:

```
//INCORRECT
int number = 0;
if( number = 0 )
{
  //some statements here
}
//CORRECT
int number = 0;
```

```
if( number == 0 )
{
    //some statements here
}
```

3. Sometimes, programmers write elseif instead of else if by mistake. Both are different, so they must be used carefully.

Switch Statement

In Java, the **switch** statement is a multi-way branching statement. It provides an easy way to forward the execution process to different parts of the program based on the value of an expression.

When one of the many alternatives is to be selected, the program can be designed using a large series of if-else-if statements to control the selection. However, the complexity of such a program increases with the number of alternatives. It becomes difficult to read, understand and follow the program. At times, it may baffle even the designer of the program. Thus, the **switch** statement often provides a better alternative than a large series of if-else-if statements.

The **switch** statement is sometimes called as a selection statement. The **switch** statement selects one option from among sections of code based on the value of an integral expression.

Integral expression is an expression that produces an integer value. The **switch** statement compares the result of integral expression to each integral value. If a match is found, the corresponding statement or statements are executed. If no match is found, the default statement is executed.

Syntax of 'switch' Statement:

```
switch (expression)
{
    case value1:
    // statement sequence
    break;
    case value2:
    // statement sequence
    break;

    ...

    case valueN:
    // statement sequence
    break;
    default:
    // default statement sequence
}
```

In the above syntax, the value of the expression is compared with each of the constant values in the case statements. If a match occurs, the code sequence following that case statement is executed else the default statement is executed. However, the default statement is optional. No further action is taken if no case matches and no default is present. Here, the expression must be of type byte, short, int, or char. The values specified in the case statements must be compatible with the type of the expression. Each value in the case statement must be a constant, not a variable. Duplicate case values are not permitted.

In the syntax, each **case** ends with a **break**. The **break** keyword results in transferring the execution to the end of the switch body. The **break** statement terminates a statement sequence. This is like "jumping out" of the switch body. Using **break** inside **switch** is optional.

 Example: Program to illustrate the use of **switch** statement in Java.

```java
class SwitchTest
{
  public static void main(String args[ ])
  {
    int month = 6;
    switch(month)
    {
      case 1:  System.out.println("January");
      break;
      case 2:  System.out.println("February");
      break;
      case 3:  System.out.println("March");
      break;
      case 4:  System.out.println("April");
      break;
      case 5:  System.out.println("May");
      break;
      case 6:  System.out.println("June");
      break;
      case 7:  System.out.println("July");
      break;
      case 8:  System.out.println("August");
      break;
      case 9:  System.out.println("September");
      break;
      case 10: System.out.println("October");
      break;
      case 11: System.out.println("November");
      break;
      case 12: System.out.println("December");
      break;
      default: System.out.println("Month not valid.");
```

```
            break;
        }
      }
    }
```
Output:

June

In this example,

1. First, a class SwitchTest is created.
2. In this class,
    (a) The main ( ) method of the class is called.
    (b) In the main ( ) method,
        (i)   An integer month is declared and assigned a value of 6.
        (ii) The value of month, that is, 6 is then compared with all the cases, one-by-one.
        (iii) Finally, the relevant case block, that is, case 6 block is executed.

3.2 Repetition Control Structures

In Java, repetition control structures or iteration statements are statements that allow a programmer to execute specific blocks of code a specific number of times. These statements are commonly called as looping statements. Java has three types of repetition control structures - the **while, do-while** and **for** loops, as given in the figure 3.2.

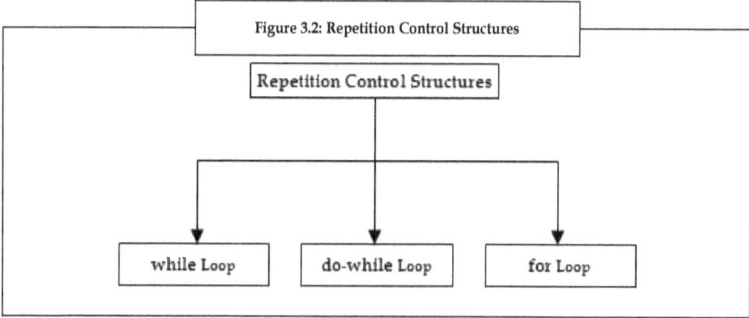

Figure 3.2: Repetition Control Structures

"while" Loop

Java's most basic looping statement is the **while** statement. This statement is an entry-controlled loop statement, which repeats a statement or block until the expression evaluates to true.

Syntax of 'while' Loop:

```
while(condition)
{
// body of loop }
```

In this syntax, **while** keyword is followed by a condition that is enclosed in parentheses and then a block of statements. This condition can be any Boolean expression. All the statements within the body of the loop will be executed as long as the conditional statement evaluates to true. The loop terminates, when the condition evaluates to false and then the control passes to the next line of code present immediately after the loop. If the condition evaluates to false, the loop is not at all executed. The curly braces are not required if only a single statement is repeated.

 Example: Program to print the table of 2 using **while** loop.

```
public class TableTest
{
  public static void main(String args[ ])
  {
        int a = 2;
        int j = 1;
        System.out.println("The table of "+a+" = ");
        while(j<=10)
        {
          int t = a * j;
          System.out.println(t);
          j++;
        }
  }
}
```

Output:

2

2

4

6

8

10

12

14

16

18

20

In this example,

1. First, a class TableTest is created.

2. In this class,

    (a) The main( ) method of the class is called.

(b) In the main( ) method,

(i)   Integer constants **a** and **j** are declared and are assigned the values of 2 and 1, respectively.

(ii) Then, the value of **a**, that is, 2 is printed.

(iii) Thereafter, the while loop starts with the condition j<=10. This condition is first checked. If it evaluates to true, the statements within the while loop are executed. Then, in the loop, for printing the table of 2, values of a and j are multiplied, and stored in the variable t. After multiplying, value of t is printed. Then, the value of j is incremented by 1to continue printing the multiplication table

Again, the condition is checked and the same process is followed till the condition evaluates to true.

Finally, the table of 2 is printed on the screen.

"do-while" Loop

The **do-while** loop is similar to the **while** loop, except that it also executes backwards. In **do-while**, the statements of the loop come before the condition, so even if the initial condition is false, the loop will execute at least once. This is the main difference between a **while** and **do-while** loop. Like the function of a **while** statement, the statements inside a **do-while** loop are executed many times as long as the condition is satisfied. The **do-while** loop is usually used, when a programmer wants to test the termination expression at the end of the loop rather than at the beginning.

Syntax of "do-while" Loop:

```
do
{
 //body of loop
}
while (condition);
```

As per the above syntax, for all the iterations of the **do-while** loop, the body of the loop first gets executed and then the conditional expression gets evaluated. If the condition evaluates to true, the loop will repeat, else the loop terminates.

Example: Program to print numbers from 0 to 5, using **do-while** loop.

```
public class DoWhileDemo
{
  public static void main(String args[ ])
  {
    int a =0;
    do
    {
      System.out.println("a is : " + a);
      a++;
```

} while(a < 6);

}

}

Output:

a is : 0

a is : 1

a is : 2

a is : 3

a is : 4

a is : 5

In this example,

1. First, a class DoWhileDemo is created.

2. In this class,

    (a) The main( ) method of the class is called.

    (b) In the main( ) method,

        (i) An integer **a** is declared, and assigned the value 0.

        (ii) Then, the do-while loop starts. In the loop, first the given value

            of **a** is printed, that is, 0. Thereafter, **a**'s value is incremented by 1.

            This process is repeated till the condition given within the while braces is true, that is, this process is repeated till the value of **a** is less than 6.

            Finally, the program gives all the values from 0 to 6 as the output.

"for" Loop

Like the previous loops, **for** loop also allows execution of the same code a set number of times. It is opposite to a **while** loop as it facilitates counting the number of iterations and stopping the loop based on this count.

Syntax of "for" Loop

for(initialization; condition; iteration)

{

    // body of the loop

}

As per the given syntax, first the initialization portion of the loop is executed. This portion of the loop includes an expression that helps in setting the value of the loop control variable. Such a variable acts as a counter, which controls the loop. The execution of this initialization expression happens only once in a program. Next, condition is evaluated, which must be a Boolean expression. This expression generally checks the loop control variable against a target value. If the result of this test is true, then the execution of the body of the loop takes place, otherwise, the loop terminates. Thereafter, the iteration portion of the loop is executed. This portion of the loop is generally an expression, which increases or decreases (increments or decrements) the loop control variable. Later, the loop iteration takes place. First the conditional expression is evaluated, then the body of the loop is executed, and then the iteration expression is executed with each pass. This process is repeated until the controlling expression evaluates to false.

Example: Program to print a list of odd numbers up to 20, using **for** loop.

```
public class TestListOddNumbers
{
    public static void main(String args[ ])
    {
        int listlimit = 20;
        System.out.println("Odd numbers between 1 and " + listlimit);
        for(int a=1; a <= listlimit; a++)
        {
            if( a % 2 != 0)
            {
                System.out.print(a + " ");
            }
        }
    }
}
```

Output:

1 3 5 7 9 11 13 15 17 19

In this example,

1. First a class TestListOddNumbers is created.

2. In this class,

   (a) The main( ) method of the class is called.

   (b) In the main( ) method,

      (i) First the limit of the list is defined, that is, the value till which this list search will proceed. For this, value 20 is assigned to listlimit. It means that odd numbers till 20 should be printed.

      (ii) Next, a statement "Odd numbers between 1 and 20" is printed.

      (iii) Then, for loop begins with the initialization of a variable **a** to 1. After initialization, a condition **a <= listlimit** is checked. Then, within for loop, if statement is used to check that the number is not divisible by 2. If this condition evaluates to true, the number is printed.

      Finally, the list of all the odd numbers between 1 and 20 are displayed on the screen as output.

3.3 Branching Statements

Just like repetition control structures, branching statements are used in Java programs. Branching statements allow the user to redirect the flow of program execution. Java offers three branching statements, as shown in the figure 3.3.

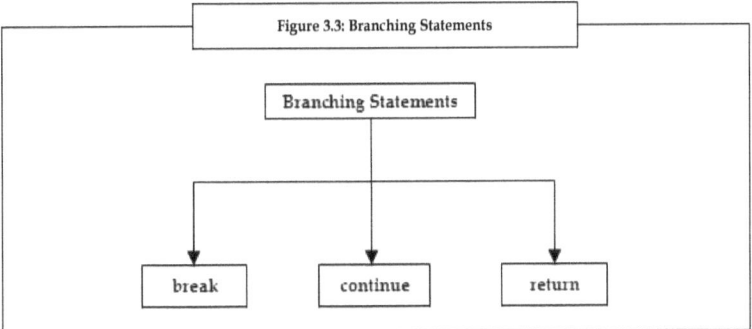

Figure 3.3: Branching Statements

"break" Statement

Break statement is one of the branching statements provided by Java, which is also used to control the flow of the program. Break statement is used to exit from a running loop program on a condition that is predefined, before the loop completion. Whenever **break** statement is used in any loop, that loop is terminated. Java programs use this to terminate the **while** loop, **do - while** loop, **for** loop and also in the **switch** statement.

Example: Program to illustrate the use of **break** statement in Java.

```java
public class JavaBreakDemo
{
    public static void main(String args[ ])
    {
        int intArray[ ] = new int[ ]{1,2,3,4,5,6};
        System.out.println("Elements less than 5 are: ");
        for(int a=0; a < intArray.length ; a ++)
        {
            if(intArray[a] == 5)
                break;
            else
                System.out.println(intArray[a]);
        }
    }
}
```

Output:

Elements less than 5 are:

1

2

3

4

In this example,

1. First, a class JavaBreakDemo is created.

2. In this class,

    (a) The main( ) method of the class is called.

    (b) In the main( ) method,

        (i) Six elements are signed to an array **intArray**. These six elements are 1, 2, 3, 4, 5, and 6.

        (ii) After this, the statement **Elements less than 5 are:** is printed on the screen.

3. Next, for loop starts with the initialization of the array element a, and proceeds with checking the condition a < intArray.length.

4. Then, if statement is used to check the condition intArray[a]==5. If it evaluates to true, the break statement breaks the loop, else the statement within the else block is executed.

This program finally gives array elements less than 5.

"continue" Statement

Just like **break** statement, **continue** statement is also a branching statement. **Continue** statement is used for stopping the execution of some statements within the loop. When **continue** statement is used, the normal flow of control is stopped, and the control returns to the loop without the execution of statements that are written after the **continue** statement. **Continue** statement can be used for skipping the present repetition of **for, while** or **do-while** loop, and begin the next repetition.

Example: Program to illustrate the use of **continue** statement in Java.

```
public class JavaContinueDemo
{
    public static void main(String args[ ])
    {
        int intArray[ ] = new int[ ]{1,2,3,4};
        System.out.println("Elements except 2 are: ");
        for(int a=0; a < intArray.length ; a ++)
        {
            if(intArray[a] == 2)
```

```
              continue;
          else
              System.out.println(intArray[a]);
          }
        }
}
```

Output:

Elements except 2 are:

1

3

4

In this example,

1. First, a class JavaContinueDemo is created.

2. In this class,

   (a) The main( ) method of the class is called.

   (b) In the main( ) method,

      (i) An array is declared with four elements (1, 2, 3, and 4).

      (ii) After declaring the array, the statement **Elements except 2 are:** is printed on the screen.

3. Next, for loop starts with the initialization of the array element a, and proceeds with checking the condition a<intArray.length.

4. Then, if statement is used to check the condition intArray[a]==2. If it evaluates to true, the continue statement skips the current iteration of the loop, else the statement within the else block is executed.

This program finally gives array elements except 2.

"return" Statement

This statement is used in the method definition, for setting the value returned by the method and for terminating the execution of the method. This means that this statement transfers the control of the program back to the caller of the method. It is also referred to as a jump statement. There are two forms of return statement, which are, one that returns a value, and one that does not return any value.

If a value is to be returned by the method, the value (or an expression that does the value calculation) is put after the **return** keyword.

 *Example:*   return "Java";

The method (in which this statement is used) returns the value Java.

If no value is to be returned by the method, that is, the method is declared void, the form of return that does not return a value is used.

Example: return;

The method does not return any value.

1. Write a program to check whether a given number is even or odd.

Lab Exercise

2. Write a program to print Fibonacci series.

3.4 Summary

- A program is defined as a group of statements, which execute in the same sequential order in which they appear, and this execution is controlled by the control constructs.

- In Java, there are three types of control constructs, namely, decision control structures, repetition control structures and branching statements.

- The decision control structures such as if, if-else, if-else-if, switch permit a programmer to select and execute particular code-blocks while bypassing other sections.

- The repetition control structures such as while, do-while, and for permit a programmer to execute particular code-blocks a number of times.

- The branching statements such as break, continue, and return permit the user to redirect the flow of program execution.

3.5 Keywords

Compound Statement: A block of statements

Constant Value: Value that does not change

Looping Statements: Executes a sequence of statements multiple times

Terminate: Come to an end

3.6 Self Assessment

1. State whether the following statements are true or false.

   (a) The return statement is used in the method definition, for setting the value returned by the method and for terminating the execution of the method.

   (b) The break statement can be used for skipping the present repetition of for, while or do-while loop, and begin the next repetition.

   (c) In do-while, the statements of the loop come before the condition, so even if the initial condition is false, the loop will execute at least once.

2. Fill in the blanks:

   (a) All _____statements evaluate to true or false for a conditional expression, to determine their execution path.

   (b) _____control structures or iteration statements are Java statements that allow a programmer to execute specific blocks of code a number of times.

   (c) Branching statements allow the user to _____the flow of program execution.

3. Select a suitable choice for every question.

   (a) Which of the following is a decision control construct?

        (i) if statement   (ii) while loop   (iii) break statement   (iv) continue statement

   (b) Which of the following is a branching statement?

        (i) if-else statement   (ii) return statement   (iii) for loop   (iv) switch statement

3.7 Review Questions

1. "All conditional statements evaluate to true or false for a conditional expression, to determine their execution path". Justify with an example.

2. "The if-statement indicates that a statement or a set of statements will be executed, if and only if a certain Boolean expression/condition is true". Discuss.

3. "When writing programs, programmers may commit few common mistakes while using the if- else statements". Discuss these mistakes.

4. "The **do-while** loop is usually used, when a programmer wants to test the termination expression at the end of the loop rather than at the beginning". Justify.

Answers: Self Assessment

1. (a) True   (b) False   (c) True

2. (a) Conditional   (b) Repetition   (c) Redirect

3. (a) if statement   (b) return statement

3.8 Further Readings

Books

Balagurusamy E. Programming with Java 3e Primer. New Delhi: Tata McGraw Publishers.

Online link

http://www.studiesinn.com/learn/Programming-Languages/Java-Language/Control-Structure.html

http://www.cs.rit.edu/~afb/20012/cs1/slides/javacontrol.html

http://cnx.org/content/m31246/latest/

http://tutorial-pemograman.blogspot.com/2009/06/strukur-control-for-java.html

Unit 4: Arrays and Strings

Objectives

After studying this unit, you will be able to:

- Define arrays

- Explain the types of arrays

- Describe strings

Introduction

Every programming language has some important concepts that make programming more easy and effective. Arrays and strings are such important concepts and are available in Java. The concept of arrays is used in a program, when similar types of values for a large number of data items are to be stored by the user. Arrays can be used for many applications such as performing calculations on statistical data and representing the state of a game.

 Example: To store the salaries of all the employees of a company, declaration of thousands of variables will be required. Also, the name of every variable must be unique. In such situations, arrays can be used for simple and easy storage.

Array is an object that contains a fixed number of values of a single data type and an array object consists of a number of variables. These variables are called as array elements. In Java, arrays can be created dynamically. The number of variables can be even zero; in such cases, the array is said to be empty. Once the array is declared, continuous space is allocated in the memory for storing these variables. The elements in the array are accessed by referring to their index value.

We can define a string as a collection of characters. Java handles character strings by using two final classes, namely, **String** class and **StringBuffer** class. The **String** class is used to implement character strings that are immutable and read-only after the creation and initialization of the string. The **StringBuffer** class is used to implement dynamic character strings.

4.1 Arrays

In an array, the memory is allocated for the same data type sequentially and is given a common name.

 Example: int salary[ ] = new int[50];

In this example, an array is created to store the salaries of 50 employees, with an array size of 50. This array is given a common name for the salaries of these 50 employees, that is, **salary.**

Since arrays are considered as objects in Java, the user must declare and instantiate them. Java arrays are of reference types. Therefore, declaration of a reference and its assignment to an available array does not mean that two objects are created, actually two references that point to the same array are created.

 Do not try to manipulate any of these objects which are created using arrays, as *Caution* any one object's manipulation will affect the other object.

4.1.1 Declaring, Creating, and Initializing an Array

Just like variables, arrays must also be declared, created, and initialized in a program, before their usage in the program.

Declaring an Array

As variables are declared before their usage in the program, so are arrays. For declaring an array, write the type of array (data type), add square brackets after the type, followed by the name of the identifier.

 *Example:* int [ ] salary;

or

int salary[ ];

Syntax of Declaring Arrays

array_var = new type[size];

In this syntax, **array_var** is the array variable that is linked to the array, **type** specifies the data type of the array, and **size** specifies the number of elements in the array.

 Example: arrayint = new int [5]; // creates an array of integers.

In this example, an array named **arrayint** is created, which stores 5 integer values in it.

An array can also be created without using the new operator as Java supports dynamic array allocation.

 Example: int[ ] arrayint = {1, 2, 3, 4, 5};

In this example, the elements of the array can be directly written within the braces. All the elements within the braces need to be of the same type.

Creating an Array

We create arrays after declaring them. We have to specify the length of the array by using a constructor that is called for the creation of a particular object. The **new** keyword is used for creating an array, which is followed by number of elements to be contained in that array (in square brackets). In any program, the array length is fixed when the array is created. We know that an array is used to represent a group of entities with the same data type in adjacent memory locations and these data items are given a common name.

 Example: students[5];

Here, the name of the array is **students**, which is of size (index value) **5**. This complete group of values is known as an array, and the individual values are called as the array elements.

 Example: int salary[ ]   // Array declaration

                  salary = new int[50]   // Array creation / Array instantiation

Initializing an Array

After the creation of an array, we need to initialize it and give a value. This means memory is allocated to that array, and this is done with the help of **new** operator, or simply at the time of declaration. If an array of integers is not initialized in the program, it will start the index numbers from 0 as it is automatically initialized to 0 at the time of creation.

Accessing an Array Element

Once the array is created and the elements are allocated in the array, a specific element of the array can be accessed by specifying its index in the square brackets. An index number is assigned to each element of the array, which allows the program and the programmer to access the individual values of the array. Index numbers are integers. These index numbers always start with zero and progress sequentially till the end of the array.

 Example: int[ ] arrayint = {1, 2, 3, 4, 5};

                  arrayint [1] = 2;

As the array index always starts from zero, the value at the index 1 of the array **arrayint** is 2.

Array Length

The Array class implicitly extends **java.lang.Object** package. Therefore, an array is an instance of Object class. An array has a named instance variable, which is called as length. The length field of an array gives the number of elements in the array. The size of the array is returned by the length field of an array.

Syntax of Array Length

            arrayName.length

In this syntax, **arrayName** specifies the name of the array and **length** specifies the length of the array.

Did you know?   We cannot resize an array after it has been declared.

4.2 Types of Arrays

Java supports different types of arrays such as one-dimensional arrays and two-dimensional arrays.

4.2.1 One-dimensional (1D) Arrays

1D array can be defined as a variables' list containing the data of similar type.

 *Example:*   A one-dimensional array can be used for storing the account numbers of the active network users.

A one-dimensional array is said to be the perfect sorting data structure because in a one-dimensional array, data is organized into an indexable linear list.

Syntax to Declare a 1D Array

type array_name [ ] = new type [size];

In this syntax, **type** specifies the type (data type) of an array, the **array_name** specifies the name of the array and size indicates the number of elements an array can store.

 Example: int salary[ ] = new int[5];

In this example, a new array of 5 elements is created using the **new** keyword, which is linked to a reference variable **salary** of an array.

Generally, a one-dimensional array can be initialized as given below:

type array-name[ ] = { val1, val2, val3, ... , valN };

In this syntax, **type** specifies the array type, **array-name** specifies the name of the array and val1 through valN specifies the initial values assigned to the array. These initial values are assigned sequentially, moving from left to right, in the order of the array indexes.

 Example: Program to illustrate the concept of 1D Array.

```
class OneDimArray
{
  public static void main(String args[ ])
  {
    int month_days[ ] = { 31, 28, 31, 30, 31, 30, 31, 31, 30, 31, 30, 31 };
    System.out.println("March has " + month_days[2] + " days.");
  }
}
```

In this example,

1. A class **OneDimArray** is created using the **class** keyword.

2. In this class, the **main( )** method of the class is called.

3. In this method,

    (a) An array **month_days** of type **int** is defined and assigned the values **{ 31, 28, 31, 30, 31, 30, 31, 31, 30, 31, 30, 31 }**.

    (b) Then, **System.out.println("March has " + month_days[2] + " days.");** statement is used to print **March has 31 days**.

When this program is executed, the number of days in **March** is printed. We know that indexes of Java arrays initiates with zero, so the number of days in **March** is month_days[2] or 31.

Control structures such as, **if** statement, **while** loop, and so on can be used in an array program to reduce the complexity of that program.

Task

Write a program to find the maximum and minimum values in an array.

4.2.2 Two-dimensional (2D) Arrays

In Java, multi-dimensional arrays are also referred to as **arrays of arrays**. 2D array is the commonly used and simplest multi-dimensional array. Generally, 2D arrays are referred to as one-dimensional arrays' lists. 2D arrays are represented in a row-column form on paper, and the terms "rows" and "columns" are used in computing.

Syntax to Declare 2D Array

type array_name = new type[rows][cols];

In this syntax, **type** specifies the type (data type) of an array, the **array_name** specifies the name of the array, **rows** and **cols** specifies the number of rows and columns in the array.

In a 2D array, memory needs to be allocated for the first dimension only and the remaining dimensions can be allocated separately.

Two-dimensional arrays can be created in two ways:

1. Reserving a block of memory that has enough space for holding all the array elements.

2. Building a multi-dimensional array from many one-dimensional arrays. (This is followed in Java).

 Example: int two-dimensional [ ] [ ] = new int [2] [3];

This statement allocates a 2 by 3 dimensional array and assigns it to two-dimensional. This 2D array is implemented as an **int** array of **int** arrays.

We can use different control structures in a 2D array program, for reducing the complexity of the program, and making it easier to read and understand.

 Example: Program to illustrate the concept of 2D Array.

```
class TwoDimArray
{
  public static void main(String args[ ])
  {
    int a, b, i;
    int table[ ][ ] = new int[4][5];
    for(a=0; a < 4; ++a)
    {
      for(b=0; b < 5; ++b)
      {
```

```
        table[a][b] = (a*5)+i+1;
        System.out.print(table[a][b] + " ");
    }
    System.out.println( );
    }
  }
}
```

In this example,

1. A class **TwoDimArray** is created using the **class** keyword.

2. In the class **TwoDimArray**, the **main( )** method of the class is called.

   Further steps are carried out in this **main( )** method.

3. In this **main( )** method, three integer variables **a, b** and **i** are declared, where **a** and **b** are used for the two dimensions of the array and **i** is used further in the program to run **for** loops.

4. Then, a two-dimensional array is created **new int [4] [5]**, which is linked to a reference variable **table** of the array.

5. After this, first **for** loop is used, wherein variable **a** is initialized to **0**, and a condition **a<4** is checked.

6. If the condition **a< 4** is true, second **for** loop starts.

7. In this **for** loop,

   (a) First, the value of **b** is initialized to **0**.

   (b) Then, the condition **b<5** is checked.

   (c) If the condition **b<5** is true, **table[a][b]** is assigned a value that comes as a result of **(a*5)+i+1;** .

   (d) Then, the **System.out.print(table[a][b] + " ");** statement is used to print the value that is assigned to **table[a][b].**

   (e) Steps (a) to (d) are repeated until the condition **b < 5** is true.

8. After the termination of the second **for** loop, **System.out.println( );** statement is used to print nothing, but to shift the cursor to the next line.

9. Steps 6 to 8 are repeated until the condition **a< 4** is true.

10. Finally, the value of **table[4][5]** will be **22** and this value is printed on screen as the output.

Write a program to multiply two matrices and print the product.

Task

4.3 Strings

Java extensively uses strings. Strings are sequence of characters or a character array. In Java, strings are objects. The Java platform uses the String class for the creation and manipulation of strings. The String class is found in the **java.lang** package, which is automatically imported.

Creating Strings

One simple form, in which the string can be created by using the **String** class is just by typing the text within the double quotes. This is called a **String** literal.

 Example: "Hello, This is Java programming."

Just like any other object, a string object can also be created using the **new** keyword and a constructor of the **String** class. The **String** class has various constructors that can be used for providing the initial value of the string using various sources like an array of characters.

 Example:

Char [ ] arraychar= {J, a, v, a}; // using the character array

String arrayofchar = new String(arraychar); // using the string object

String Length

We can calculate the length of the string by using the length method, which returns the number of characters in the string.

Syntax of String Length:

        String1.length( );

In this syntax, **String1** is the name of the string in which the **length( )** method is called.

 Example: Program to count the number of characters in the string.

```
import java.lang.*;

public class StringLen

{

    public static void main(String args[ ])

    {

        String str = "Programming"; int length =

        str.length( ); System.out.println("String length

        is: " + length);

    }

}
```

Output:

The length of the string is: 11

In this example,

1. First the **java.lang.*** package is imported, using the **import** keyword, for strings.

2. Then, a class **StringLen** is created and declared public with the **public** keyword.

3. In the class **StringLen, main( )** method of the class is called, using the **public static void main(String args[ ])** statement.

4. In this method,

(a) First, a **String str** is declared for holding the string and is assigned a string **Welcome**.

(b) Then, integer **length** is assigned the **str.length( );** method.

(c) Finally, the **System.out.println("The length of the string is: " + length);** statement is used to print the length of the string, that is, the number of characters in a string. As **Programming** consists of 11 characters, the output will be:

String length is:11

4.3.1 String Array

As the name suggests, string array is an array containing strings. We can declare string arrays in the following two ways:

1. With an initial size

2. Without an initial size

1. *With an Initial Size:* In Java, we can declare string arrays and assign an initial size to them.

Example:
```
public class JavaStringArrayDemo
{
    private String[ ] fruits = new String[10];
    // more to the class here ...
}
void populateStringArray( )
{
    fruits[0] = "Apple";
    fruits[1] = "Mango";
    fruits[2] = "Banana";
    // ...
}
```

In this example,

1. A class **JavaStringArrayDemo** is created, and declared public using the **public** keyword.

2. In the class **JavaStringArrayDemo**, a **String** array named as **fruits** is created, where the **fruits** array has been given an initial size of **10** elements.

3. Then, the elements in the String array are assigned by the **populateStringArray( )** method in the class:

(a) **fruits[0]** is assigned a string **Apple**.

(b) **fruits[1]** is assigned a string **Mango**.

(c) **fruits[1]** is assigned a string **Banana**.

Did you know?   A String array in Java begins with an element numbered zero.

2. **Without an Initial Size:** We can also declare a Java String array without giving it an initial size.

Example: public class JavaStringArrayDemo

{

private String[ ] toppings;

// more to the class here ...

}

After this, Java array can be given a size in the program code, and populated as desired, like this:

void populateStringArray( )

{

fruits[0] = "Apple";

fruits[1] = "Mango";

fruits[2] = "Banana";

// ...

}

This method of declaring an array is very similar to the first method. However, in this method, the string array is not given any size until the **populateStringArray** method is called.

4.3.2   String Methods

A number of methods are defined in the String class. These methods are called as String methods, which are used for different tasks of string manipulation. These string methods and the manipulation tasks performed by them are as follows:

1. **str2 = str1.toLowerCase; :** This method is used to change the string **str1** to all lowercase and assigns the value to **str2**.

2. **str2 = str1.toUpperCase; :** This method is used to change the string **str1** to all uppercase and assigns the value to **str2**.

3. **str1.concat (str2) :** This method is used to concatenate **str1** and **str2**.

4. **a.toString( ) :** This method is used to create object **a**'s string representation.

5. **str1.length( ) :** This method is used to get the length of **str1**.

6. **str1.substring (n):** This method is used to get a substring that begins from the nth character.

7. **str1.substring(n, m) :** This method is used to get a substring that begins from the nth character till the mth character.

8. **str2 = str1.replace('a', 'b'); :** This method is used to replace all the appearances of **a** with **b**.

9. **str2 = str1.trim ( ); :** This method is used to remove the white spaces (if any) from the starting and the end of the **str1**.

10. **str1.equals (str2) :** This method is used to return **true**, when **str1** is equal to **str2**.

11. **str1.compareTo(str2)** : This method is used for comparing two strings. If **str1 < str2**, it returns negative, and if **str1 > str2**, it returns positive. If **str1 = str2**, then it returns 0 (zero).

12. **str1.append (str2)** : This method is used to append **str1** to **str2** at the end.

13. **str1.setLength(a)** : This method is used for setting the length of **str1** to **a**.

14. **str1.insert (a, str2)** : This method is used for inserting the **str2** at position **a** of **str1**.

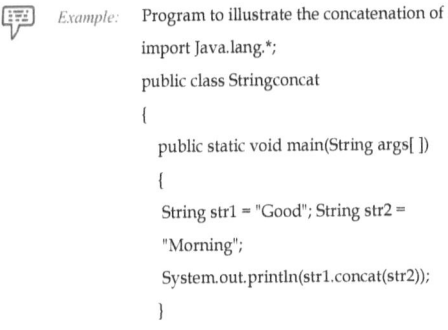

Example: Program to illustrate the concatenation of strings using the **concat( )** method.

```
import Java.lang.*;
public class Stringconcat
{
    public static void main(String args[ ])
    {
    String str1 = "Good"; String str2 =
    "Morning";
    System.out.println(str1.concat(str2));
    }
}
```

Output:

Good Morning

In this example,

1. A class **Stringconcat** is created, and declared public with the **public** keyword.

2. In the **Stringconcat** class, the **main( )** method of the class is called using the **public static void main(String args[ ])** statement.

3. In the main( ) method,

    (a) Two string variables **str1** and **str2** are declared, and assigned the strings **Good** and **Morning** respectively.

    (b) Then, in the **System.out.println(str1.concat(str2));** statement, **concat( )** method is called for the concatenation of both **str2** and **str2** and print the output: **Good Morning**

4.3.3 String Operators

As arithmetic operators are used in arithmetic expressions, string operators are used for string operations. Let us next discuss the different types of string operators.

== Operator

This operator is used for comparing the references to string objects. This comparison returns **true**, only if these two string variables indicate the same object in the memory, else it returns **false**.

Caution

The == operator cannot be used for the comparison between the content of the text present in the string objects, as it only compares the references that the two strings are pointing to.

Example: Program to illustrate the use of == operator to compare two strings.

```
public class CompareStrings
{
public static void main(String args[ ])
{
    String cellphonename1 = "Nokia";
    String cellphonename2 = "Nokia";
// 1st case
if (cellphonename1.equals(cellphonename2))
    {
        System.out.println("Equal Strings");
    }
else
    {
        System.out.println("Unequal Strings");
    }
// 2nd case
if (cellphonename1.==cellphonename2)
{
        System.out.println("Equal Strings");
}
else
{
        System.out.println("Unequal Strings");
    }
}
} Output:
Equal Strings
Unequal Strings
```

In this example,

1. A class **CompareStrings** is created and declared public with the **public** keyword.

2. In the **CompareStrings** class, **main( )** method of the class is called in the **public static void main(String args[ ])** statement.

3. In this method,

   (a) A **String cellphonename1** is declared and assigned the string **Nokia**.

   (b) Then, one more string **String cellphonename2** is declared and is also assigned the string **Nokia**.

   //For 1st Case

   (a) The **if** statement is used to compare **cellphonename1** and **cellphonename2** to check whether they are equal or not, using the **if (cellphonename1.equals(cellphonename2))** statement.

   (b) If the condition **(cellphonename1.equals(cellphonename2))** is true, **Equal Strings** is printed on the screen as the output, but if the condition is false, **Unequal Strings** is printed on the screen as the output. As the contents of **cellphonename1** and **cellphonename2** are the same, **Equal Strings** is printed on the screen as the output.

   //For 2nd Case

   (a) The **if** statement is used to compare the references of **cellphonename1** and **cellphonename2** to check whether they are equal or not, using the **if (cellphonename1 == cellphonename2)** statement.

   (b) If the condition **(cellphonename1==cellphonename2)** is true, **Equal Strings** is printed on the screen as the output, but if the condition is false, **Unequal Strings** is printed on the screen as the output. As the references of **cellphonename1** and **cellphonename2** are not the same, **Unequal Strings** is printed on the screen as the output.

+ Operator

This operator can be used to concatenate two strings.

Syntax for Using + Operator

string 3 = string 1 + string 2;

 Example: Program to illustrate the concatenation of strings using + operator.

```
import java.lang.*;
public class ConcatStringDemo
{
  public static void main(String args[ ])
  {
    String str1 = "My name is";
    String str2 = "John";
    String str3 = str1 + " " + str2;
```

NAGESH JAITAK

```
System.out.println(str3);

    }
}
```

Output:

My name is John

In this example,

1. A class ConcatStringDemo is created, and declared public using the public keyword.

2. In the class ConcatStringDemo, the main( ) method of the class is called using the public static void main(String args[ ]) statement.

3. In this method,

    (a) Two strings **str1**, and **str2** are declared and assigned the strings **My name is** and **John** respectively.

    (b) A third string **str3** is declared, which will store the output of concatenation of **str1** and **str2**. This concatenation is done using the + operator.

4. Finally, the System.out.println(str3); is used to print the value stored in **str3**, that is, **My name is John**.

4.3.4   StringTokenizer

Java packages consist of many classes. **StringTokenizer** class is one such class, which exists in **java.util.package**. This class is mainly used for parsing a string into tokens by certain delimiters.

The objects which are created from the **StringTokenizer** class represent the StringTokenizers. Any of the following three constructors of **StringTokenizer** class is called to create these objects.

1. *StringTokenizer st = new StringTokenizer(s); :* This constructor is used to construct a StringTokenizer **st** for the string **s**, in which, whitespaces are used as delimiters.

2. *StringTokenizer st = new StringTokenizer(s, a); :* This constructor is used to construct a StringTokenizer **st** for the string **s** using delimiters from the string **a**.

3. *StringTokenizer st = new StringTokenizer(s, a, b); :* This constructor is used to construct a StringTokenizer **st** for the string **s** using delimiters from the string **a**. If the Boolean **b** is **true**, each delimiter character will also be returned as a token.

Notes   The constructor returning false specifies that tokens are not returned.

All these constructors take argument(s), that is, a reference to a string. This argument specifies the string that has to be tokenized.

After the creation of the tokenizer objects, different methods of **StringTokenizer** class can be called for counting the number of tokens, checking whether more tokens are available or not, and returning a token. These methods are described below.

1. *StringTokenizer.hasMoreTokens( ):* This method returns **true** if there are more tokens available in the string.

2. ***StringTokenizer.nextToken( ):*** This method returns the token that comes immediately after the present token, as a String.

3. ***StringTokenizer.countTokens( ):*** This method returns the total number of tokens.

4.3.5 Exploring the String Class

There are many classes in the class library of Java. String class is the most widely used class in this library, because strings are very important in Java programming. Every string that is created by the user is a **String** type object. In Java, String constants are handled in the same manner as the normal strings are handled.

String Class Constructors

The **String** class consists of various constructors to create objects of **String** type. These constructors are:

1. ***String( ):*** This constructor is the default constructor of the **String** class, which is used to create a **String** object without any value or an empty object of **String** type.

2. ***String(char arr[ ]):*** This constructor is used to initialize the object of the **String** class by assigning a value to it.

 Example: char arr[ ] = {'J', 'A', 'V', 'A'}

   String obj = new String(arr);

   In this example, the array **arr** is passed as an argument to the constructor. The object **obj** of the String class is assigned an initial value, that is, the string "JAVA".

3. ***String(char arr[ ], int start, int length):*** This constructor is used to assign a string to an array containing only the array's sub-array.

 Example: char arr[ ] = {'J', 'A', 'V', 'A'}

   String obj = new String(arr, 1, 2);

   In this example, the array **arr** is passed as an argument to the constructor along with the starting position of the array and the number of elements. The object **obj** of the String class is assigned an initial value, that is, the first two characters of the string "JAVA", which is 'AV'.

4. ***String(String obj):*** This constructor is used to assign the same string value to the new string object, as the String object **obj** has.

Once a String object is created, the user cannot change its contents. This could seem to be a serious restriction, but it is not, because:

1. If a string needs to be changed, the user can create a new string containing the alterations.

2. In Java, a peer class of string is defined, which is called as **StringBuffer**. This class permits the alteration of strings.

String Buffer Constructors

The **StringBuffer** class provides three types of constructors.

1. ***StringBuffer( ):*** This constructor is the default constructor of the **StringBuffer** class, which does not contain any parameters. This constructor first constructs an object and then does its initialization with no character sequence. It has the capacity to store 16 characters.

Example: StringBuffer sb1 = new StringBuffer( );

> In this example, **StringBuffer** is the name of the class, and **sb1** is an object of that class, which is initialized with no character sequence. The object **sb1** has the capacity to store 16 characters.

2. ***StringBuffer(int val):*** This constructor takes an argument of integer type, which is used to set the buffer size. This constructor first constructs an object and then does its initialization with no character sequence. It has the capacity to store the number of characters that is specified by the value of variable **val**.

 Example: StringBuffer sb1 = new StringBuffer(10);

> In this example, **StringBuffer** is the name of the class and **sb1** is an object of that class, which is initialized with no character sequence. The object **sb1** has the capacity to store 10 (this value is passed as an argument) characters.

3. ***StringBuffer(String obj):*** This constructor accepts the string object **obj** as an argument. This constructor first constructs an object and then does its initialization with the character sequence which is similar to the character sequence of string object **obj**. It has the capacity to store the number of characters of the string object **obj** and 16 more characters additionally.

 Example: StringBuffer sb1 = new StringBuffer("Java");

> In this example, **StringBuffer** is the name of the class and **sb1** is an object of that class, which is initialized with the **Java** character sequence. The object has the
>
> capacity to store the number of characters of the string **Java** and 16 characters additionally.

StringBuffer Methods

The **StringBuffer** class has some methods that can be used to manipulate the objects of that class.

1. ***length( ):*** This method is used to return the length of an object, that is, the number of characters that are contained in an object.

 Example: StringBuffer sb1 = new StringBuffer("Java");

> int a = sb1.length( );
>
> In this example, **StringBuffer** is the name of the class and **sb1** is an object of that class, which is initialized with the **Java** character sequence. Then, an integer variable **a** is declared, which stores the length of the object. The **length( )** method is used to find the length of the string **sb1**, which gives 4 as output.

2. ***capacity( ):*** This method is used to return the number of characters that an object can store without any increase in the object's capacity. Syntax of this method:

> int capacity( )

Example: StringBuffer sb1 = new StringBuffer("Java");

int a = sb1.capacity( );

In this example, **StringBuffer** is the name of the class, and **sb1** is an object of that class, which is initialized with the **Java** character sequence. Then, an integer variable **a** is declared, which stores the capacity of the object. The **capacity( )** method is used to find out the capacity of the object **sb1**, which finally gives 20 as output. Out of these 20 characters, 4 characters are of **Java**, and remaining 16 are already reserved.

These methods help in the manipulation of the objects of the **StringBuffer** class.

Lab Exercise

1. Write a program to print the sum of two matrices.
2. Write a program to swap the values of two arrays, and print the output.

4.4 Summary

- A Java array is an object containing fixed number of values of a single data type, which can be created dynamically and is given a common name.
- In a Java program, the length of an array is fixed and it starts with zero (0).
- An array can be created with or without using the **new** operator.
- Initialization of an array can be done at the time of array declaration, or after the array creation.
- Array length specifies the number of array elements in an array.
- One-dimensional arrays are defined as a variables' list comprising the data of same type.
- Generally, two dimensional or 2D arrays are referred to as one-dimensional arrays' lists. 2D arrays are represented in a row-column form
- A string is a sequence of characters which is created using the **String** class.
- The length of a string refers to the number of characters in a string.
- An array that contains strings is a string array.
- String class consists of some methods for creating string objects. These methods are called as **String** methods.
- Two operators are used for **String** operations, = = operator, and + operator.
- The **java.util.package** consists of a class named as **StringTokenizer** class, which is mainly used for parsing a string into tokens by certain delimiters.
- The **StringBuffer** class is a peer class of String, which is used for strings' alteration.

4.5 Keywords

Arguments: A value that is passed to a function, procedure, subroutine, command, or program.

Delimiters: Strings that point to the separation between tokens.

Memory Location: A memory address where data in a computer program is stored.

Parsing: Division of string text into different parts, which are known as tokens, used to convey a semantic meaning.

NAGESH JAITAK

4.6 Self Assessment

1. State whether the following statements are true or false:

    (a) Java arrays can be created dynamically.

    (b) The elements in the array are accessed by referring to their index value.

    (c) The **==** operator is used for the comparison between the content of the text present in the String objects.

    (d) The **StringTokenizer** class is mainly used for parsing a string into tokens by certain delimiters.

    (e) An array cannot be created without using the new operator.

    (f) A String array in Java begins with an element numbered zero.

2. Fill in the blanks:

    (a) The Array class implicitly extends _____package.

    (b) _____method is used to return true, if there are more tokens available in the string.

    (c) _____method is used to append str1 to str2 at the end.

    (d) The length of the string can be calculated by using the length method, which returns the number of _____in the string.

    (e) The Java platform uses the String class for the creation and manipulation of _____.

3. Select a suitable choice for every question.

    (a) Which of the following methods is used to create object a's string representation?

        (i)   a.toString( )

        (ii) StringTokenizer.countTokens( )

        (iii) str1.append (str2)

        (iv) str2 = str1.replace('a', 'b');

    (b) Which of the following methods is used to return the token that comes immediately after the present token, as a String?

        (i)   StringTokenizer.countTokens( )

        (ii) StringTokenizer.nextToken( )

        (iii) StringTokenizer st = new StringTokenizer(s, a, b);

        (iv) StringTokenizer st = new StringTokenizer(s, a);

    (c) Which constructor is used for initializing the object of the String class by assigning value to it?

        (i)   String(char arr[ ])

        (ii) StringBuffer(String obj)

        (iii) StringBuffer( )

        (iv) String(char arr[ ], int start, int length)

    (d) Which method is used to return true, if there are more tokens available in the string?

        (i) StringTokenizer.hasMoreTokens( )   (ii) StringTokenizer.countTokens( )

        (iii) capacity( )   (iv) str1.length( )

(e) Which method is used to remove the white spaces (if any) from the starting and the end of the str1?

(i) str1.append (str2)   (ii) str2 = str1.trim( );   (iii) str1.insert (a, str2)   (iv) a.toString( )

4.7 Review Questions

1. "In an array, the memory is allocated for the same data type sequentially and is given a common name." Justify.

2. "An array can also be created without using the new operator as Java supports dynamic array allocation." Justify with an example.

3. "In any program, the array length is fixed, when the array is created." Comment.

4. "After the creation of an array, it should be initialized and given a value." Discuss.

5. "Size of the array is returned by the length field of an array." Discuss.

6. "2-dimensional arrays can be created in two ways." Justify.

7. "String arrays can be declared with an initial size, and without an initial size." Do you agree? Justify with examples.

8. "Various string methods are used for different tasks of string manipulation." Discuss these methods.

9. "As arithmetic operators are used in arithmetic expressions, similarly string operators are used for string operations." Discuss these string operators.

10. "After the creation of the tokenizer objects, different methods of **StringTokenizer** class can be called for counting the number of tokens, checking whether more tokens are available or not, and returning a token." Elaborate.

11. "The **StringBuffer** class consists of some methods that can be used for the manipulation of the objects of that class." Elaborate.

12. "Once a String object is created, the user cannot change its contents. This seems to be a serious restriction, but it is not so." Why?

Answers: Self Assessment

1. (a) True   (b) True   (c) False   (d) True   (e) False   (f) True

2. (a) java.lang.Object   (b) StringTokenizer.hasMoreTokens( )   (c) str1.append(str2)

   (d) Characters   (e) Strings

3. (a) a.toString( )   (b) StringTokenizer.nextToken( )   (c) String(char arr[ ])

   (d) StringTokenizer.hasMoreTokens( )   (e) str2 = str1.trim( );

4.8 Further Readings

Books

Balagurusamy E. Programming with Java 3e Primer. New Delhi: Tata McGraw Publishers.

Schildt H. Java A Beginner's Guide, 3rd ed. New York: Mc-Graw Hill

Online link

http://www.javabeginner.com/learn-java/java-string-comparison

http://www.leepoint.net/notes-java/data/strings/55stringTokenizer/10stringtokenizer.html

http://admashmc.com/main/images/Lec_Notes/javaarray.pdf

Unit 5: Packages

Objectives

After studying this unit, you will be able to:

- Describe packages
- Discuss the importance of access protection
- Explain the method used to import packages
- Identify the standard Java packages

Introduction

Java OOP has one important feature which states that the code that is already created can be reused by extending the classes and implementing the interfaces. This ability to reuse the code is limited only to the classes that exist within a program. To use the classes that are present in other programs, we can use Java packages. While using Java packages, the classes are not copied physically into the program being developed. In Java, a method, by which the class name space can be partitioned into more manageable chunks, is known as a **package**.

Packages are similar to the header files used in C and C++. Packages are the library files in Java. They are a collection of similar classes and interfaces. Java has many in-built packages. Package provides both naming and visibility control mechanism. Code outside a package cannot access classes that are defined within the package. We can define class members in such a way that these class members are exposed only to other members of the same package. This property of packages allows the classes to have knowledge of each other, but not share the knowledge.

 Example: Java allows you to create a class named Test, which can be stored in your own package. This Test package will not collide with any other class named Test stored elsewhere.

The two main reasons for using packages are:

1. To avoid conflicts among classes

2. To facilitate reusability of code

If there are some applications that need the same code or method, then it is better to have the common code in a separate file, and just use it in the application, wherever needed. Reusability of code is possible in interface and inheritance as well, but putting the code separately in a package makes it easy to manage the reusability of code in the classes.

Usually, packages are stored in a corresponding folder in the file system, but packages can also be stored in a database. The name of the folder in the file system must have the same name as that of the package. The folder has all the classes which belong to the package.

5.1 Defining a Package

Defining a package is simple. Just include a package command in the first statement in a Java source file. All the classes declared within that file belong only to the specified package. The **package** statement basically describes a name space that contains all the classes. The **package** statement specifies the package, to which, the classes defined in a file belong. Suppose **package** statement is not used, all the class names will be put into the default package, which has no name. Default packages suffice for short and simple programs, but for real applications, it is inadequate. It is a good practice to define a package for your code.

General form of the **package** statement:

> package pkg;

Here, **package** is the keyword, and **pkg** is the name of the package.

 Example: package ExamplePackage;

> In this example, **package** is the keyword, and **ExamplePackage** is the name of the package that is created.

In Java, packages are stored in file system directories.

Notes   The class files for any classes that are declared as a part of **ExamplePackage** must be stored in a directory called **ExamplePackage**.

Multiple files can include the same **package** statement.

Caution   Java is case-sensitive. Hence, the directory name must match the package name exactly.

It is possible to create a hierarchy of packages. It is done by using a period in between the names of packages.

General form of a hierarchical package statement:

> package pkg1[.pkg2[.pkg3]];

 Example: The standard Java packages are best examples for hierarchical packages. One such package is java.util.

A hierarchical package must be added in the file system of the Java development system.

Example. A package called package **java.awt.font** must be stored in **java/awt/font, java\awt\font,** or **java:awt:font** on the UNIX, Windows, or Macintosh file system, respectively.

The two basic requirements for creating a package are:

1. There should be one or more interfaces or classes in a package. This means that the package cannot be empty.

2. The source code of the interfaces or classes that the package contains should be in the same directory structure, as named in the package.

Having packages in a programming language is important, as it makes the work of the programmer easy by holding all the related files in the same package. There are many library files or packages that are in-built or predefined in Java. On the other hand, the users can create their own packages, which are called as the user-defined packages. Thus, it can be said that there are two types of packages in Java, which are:

1. Predefined Packages

2. User-defined Packages

Predefined Packages

As the name suggests, predefined packages are the packages that are already defined. They are the in-built packages or the library files. There are many predefined classes and interfaces in Java that are used while developing an application. These classes and interfaces are arranged in groups, and added to a folder in the library files. Thus, they are called as predefined packages.

 Example: The classes that are needed for the input/output operations, are put inside the java.io package. Similarly, the classes that are required for creating applets are put under the **java.applet** package, and so on. This helps the java compiler to search for the class in that particular package. Suppose an application is written to demonstrate the use of input/output operations, then the **java.io** package is imported in the application. This helps the java compiler to search for the input
/output streams in a particular class in the **java.io** package.

Whenever there is a need for any package in a class, it can be imported to the class by just using the **import** keyword.

 Example: import java.io.*

If the input/output files are needed in an application, then it can be imported by mentioning import java.io.* at the beginning of the application. This imports all the classes in the io package.

General form of predefined packages:

> import java.[package name].*;

 Example: import java.string.*;

The above code imports all the classes in the string class, which has all the string operations.

User-defined Packages

Another type of package supported by Java is the **user-defined package**. In Java, programmers can define their own packages, if there are similar classes and the classes are needed in more than one application.

Creating a User-defined Package

When a package is created, you have to first identify the classes and interfaces which can be included that package.

While creating a package, first create a folder in the name of the package.

 Example: If the name of the package is **Mypack,** then name the folder as **Mypack** and copy all the files, which belong to the **Mypack** package to this folder.

While importing the package in any file, add the package name at the top of the class file.

Naming Packages

While naming a package, first specify the domain name (if there is one) and then add the project name. If there is no domain name, then the common way of naming the package is used, which is same as discussed earlier. Even a user-defined package is created using the syntax as given below.

The general form of naming a package (or user-defined package) is:

> package pkg;

As per this syntax, while naming a package first write the keyword **package** and then the name of the package.

 Example: Program to illustrate the usage of packages in Java.

```
package MyPack;
class Emp
{
String name;
int employeeid;

Emp(String n, int empid)
{
name = n;
employeeid = empid;
}

public void display( )
{
System.out.println("Employee: " + name);
System.out.println("Employee Id: " + employeeid);
}
```

```
}
class EmployeeDetails
{
public static void main(String args[ ])
{
  int p;
EMP e = new EMP(4);
e[0] = new EMP("Ax", 101);
e[1] = new EMP("By", 102);
e[2] = new EMP("Cz", 103);
e[3] = new EMP("Dk", 104);

for(p = 0, p < 4, p++) e[p].display( );
}
}
```

Output:

Employee: Ax

Employee Id: 101

Employee: By

Employee Id: 102

Employee: Cz

Employee Id: 103

Employee: Dk

Employee Id: 104

In this example,

1. First a user-defined **package Mypack** is defined.

2. Then, a class **Emp** is declared.

3. In this class **Emp**,

    (a) A string **name**, and an integer **employeeid** are declared.

    (b) Then, a constructor **Emp(String n, int empid)** is called, wherein:

      (i) String **n** is assigned to **name**.

      (ii) Integer empid is assigned to employeeid.

      (iii) Then, the **display( )** method is called for displaying the below-given output on the screen.

      (iv) First, **System.out.println("Employee: " + name);** statement is used to print the name of the employee on the screen.

(v) Then, **System.out.println("Employee Id: " + employeeid);** statement is used to print the employee id on the screen.

4. A new class **EmployeeDetails** is created.

5. In this class EmployeeDetails,

   (a) The main( ) method of the class is called.

   (b) In this main( ) method,

      (i) An integer **p** is declared.

      (ii) A new employee EMP is created with the parameter 4, by using the new keyword, and is linked to EMP e.

      (iii) Then, new employees EMP are created with parameters name and employeeid, and are linked to employees e[0] to e[3], respectively.

      (iv) In for loop, value of **p** is initialized with 0, and **p** < 4 condition is checked. If this condition is true, then the respective employee's name and id e[0] will be displayed, that is, e[p].display( ); After printing this, the value of **p** is incremented by 1. This process continues till the condition is true.

Note: In this example, the employee names are given as Ax, By, Cz, Dk, and the employee IDs are 101, 102, 103, and 104.

Did you know?   Many Java packages work with Really Simple Syndication (RSS). RSS is a method by which regularly changing Web content is delivered to the user's browser or desktop.

Adding Class to a Package

After the package is created, it is stored in the folder with the name of the package. Then, classes and interfaces are added into that folder.

Adding a class to an already existing package is done by adding the statement with the **classname.java** source file in the first line of the application.

General Form of Adding a Class to a Package:

   [Class name].java;

package [package name];

In this syntax of adding a class to a package, the first statement specifies the class, which has to be added to the package. This statement will be the first statement in the application. In the second statement, the keyword **package** is used along with the name of the package, to define the package.

 Example: One.java;

   package Mypack;

   In this example, a class **One** is added to the package **Mypack**. To add the class **One.java**, a statement with the class **name.java** that is 'One.java' is added as the first statement.

Classes can be hidden in such a way that some important part of the code is blocked, so that the client cannot access it.

5.2 Finding Packages and CLASSPATH

From the above explanation, it is quite clear that packages are known with their respective directories. This raises a question. How does Java run-time system know where to look for a package that was created by a user and is used in a program?

This question has two answers:

1. Java run-time system always uses the current working directory as its starting point. Therefore, if the user-defined package is present in the current directory or the sub-directory of the current directory, it will be found.

2. The user can also specify a directory path or paths by setting the CLASSPATH environmental variable.

 Example: Consider the following package specification.

package ExPack;

If a program has to find **ExPack,** either of the two things must be true – the program is executed from a directory which is just above **ExPack** or CLASSPATH must be set to include the path to **ExPack.**

The first alternative provided in the above example is easier as it does not require a change in the CLASSPATH. However, the second alternative allows your program to find **ExPack,** whichever directory the program is placed in. Ultimately, the programmer decides the method to be used in the program.

 Example: Program to illustrate the usage of CLASSPATH.

```
package AccountPack;
class MyBalance
{
String name; double balance;
MyBalance(String n, double b)
{
   name = n;
   balance = b;
}
void show( )
{
   if(balance<0) System.out.print("--> ");
   System.out.println(name + ": $" + balance);
}
}
class MyAccountBalance
```

```
{
    public static void main(String args[ ])
    {
        MyBalance current[ ] = new MyBalance[3];
        current[0] = new MyBalance("Sanjay", 333.45);
        current[1] = new MyBalance("Sharath", 138.91);
        current[2] = new MyBalance("Asha", 245.67);
        for(int i=0; i<3; i++) current[i].show( );
    }
}
```

Name this file as **MyAccountBalance.java** and put it in a directory called **AccountPack**.

The next step is to compile the file. While compiling, make sure that the resulting **.class** file is also in the **AccountPack** directory. Then execute the **MyAccountBalance** class, using the following command line:

java AccountPack.MyAccountBalance

It is essential that you are in the directory above AccountPack when you execute this command, or you have your CLASSPATH environmental variable set appropriately. As explained, MyAccountBalance is now part of the package AccountPack. Hence, it cannot be executed by itself. Therefore, MyAccountBalance must be qualified with its package name.

Advantages of Using a Package

The advantages of using a package are:

1. It helps in finding the classes and interfaces easily, as they are grouped together in a single and specific package.

2. It helps in having more than one class with the same name when their functionality is same, but they should be in different package.

Did you know?   Java packages are stored in a file called Java Archive (JAR) file. Packages stored in a JAR file can be optionally sealed in such a manner that the package can implement version consistency.

5.3 Access Protection

We know that hiding of classes is possible in Java. It is done using the access control keywords - **public**, **private** and **protected**.

1. *Public Modifier:* When a class, a method and the variables are declared as **public**, these methods, classes and variables are visible from the subclasses in the same package, subclasses in the other packages, non-subclass in the same package and non-subclass in the other packages.

2. *Protected Modifier:* When a class, a method and the variables are declared as **protected**, these methods, classes and variables are visible from the subclasses in the same package, subclasses in the other packages, non-subclass in the same package and are not visible from the non-subclass in the other packages.

3. **Default Modifier.** When there is no access modifier specified for a class, a method, or variable, then the methods, classes and variables are considered as **default.** These methods, classes and variables are visible from the subclasses in the same package and non-subclass in the same package, whereas they are not visible from the subclasses in the other packages and the non- subclass in the other packages.

4. **Private Modifier.** When a class, a method and the variables are declared as **private,** these methods, classes and variables are not visible from any class or subclass outside the class.

Packages add a different dimension to access control. Java provides multiple levels of protection, which facilitates fine-grained control over the visibility of variables and methods within packages. The main purpose of access protection is to protect variables from external modification.

Access modifiers that are used in classes, block some important part of the code in such a manner that this block becomes invisible to the users. Similarly, there are some access modifiers in packages also, which are used to hide some classes of the package in the main application.

Classes and packages are used to encapsulate the variables and methods of the classes, and limit the scope of the variables. Packages are containers for classes, and classes are containers for data and code. As the packages and classes have some kind of inter-relationship, the packages in Java are categorized into four different types based on their relationship with each other. These categories are:

1. Subclasses that are in the same package

2. Subclasses that are in different packages

3. Non-subclasses that are in the same package

4. Classes that are neither found in the same package nor in the subclasses

Table 5.1 illustrates the access specification in packages.

Table 5.1: Access Modifiers in Packages

Categories of the package	Public	Protected	Default	Private
Subclasses that are in the same package	Yes	Yes	Yes	No
Subclasses that are in different packages	Yes	Yes	No	No
Non-subclasses that are in the same package	Yes	Yes	Yes	No
Non-subclasses that are in different packages	Yes	No	No	No

Benefits of Access Protection

Access protection provides the following benefits:

1. It permits the enforcement of constraints on an object's state.

2. It provides a simpler client interface. Client programmers do not need to know everything that is present within the class. It is sufficient if they know what is present in the public parts of a class.

3. It separates interface from implementation, allowing them to vary independently.

Example: For instance consider making the licensePlate field of Car an instance of a new LicensePlate class instead of a **String.**

5.4 Importing a Package

Once the package is created and executed, it can be used in different classes. To use a package in any class, the package has to be imported in the class, which has to use it. Importing the package can be done in three ways, which are:

1. Using the fully qualified name of a package

2. Importing the package and the class

3. Importing the package completely

Using the Fully Qualified Name of a Package

In this method of importing a package, the package is imported till the method level.

General Form of Fully Qualified Name of a Package

>   Package [package name].[class name].[method];

In the syntax, **package** is the keyword and the **package name** is the name of the package to which the method is being imported, **class name** is the name of the class to which the method belongs, and the **method** is the name of the method that is being imported.

 Example: package Mypack.EMP.display( );

>   In this example, the package **Mypack** is imported till the method level. Here, the package **Mypack** has a class **EMP,** which has a method **display( ).** The package **Mypack** is imported till the method **display( )** of the class **EMP.**

When packages are imported in this way, only the specified method (of the package) will be available to the class.

Importing the Package and the Class

In this method of importing a package, the package is imported till the class level.

General Form of Importing the Class of a Package

>   package [package name].[class name];

In the syntax, **package** is the keyword, **package name** is the name of the package to which the class is being imported and the **class name** is the name of the class which is being imported.

 Example: package Mypack.EMP;

>   In this example, the package **Mypack** is imported till the class **EMP** level.

Importing the Package Completely

In this method of importing a package, the package is imported till the package level, that is, all the classes and interfaces of the package are imported to the class, which uses the package.

General Form of Importing a Package Completely

>   package [package name].*;

Here, **package** is the keyword and **package name** is the name of the package that has to be imported. In the above syntax, the * at the end indicates that all the classes and interfaces of the package are imported into the class, which imports it.

NAGESH JAITAK

Example: package Mypack.*;

In this example the package **Mypack** is imported with all its classes and interfaces till the package level.

Caution The * used in the above syntax may increase the compilation time. This happens most of the time, in those cases, where several large packages are imported. In such cases, it is better to explicitly name the classes that you use, instead of importing whole packages. However, it is important to note that the * has no effect on the run-time performance or size of the classes in a program.

5.5 Basics of Standard Java Packages

Java Application Programming Interface (API) facilitates the grouping of a large number of classes into different packages. These classes are grouped into packages based on their functionality. Mostly, programmers use the packages that are available with the Java API.

Figure 5.1 illustrates some of the packages that are frequently used in Java.

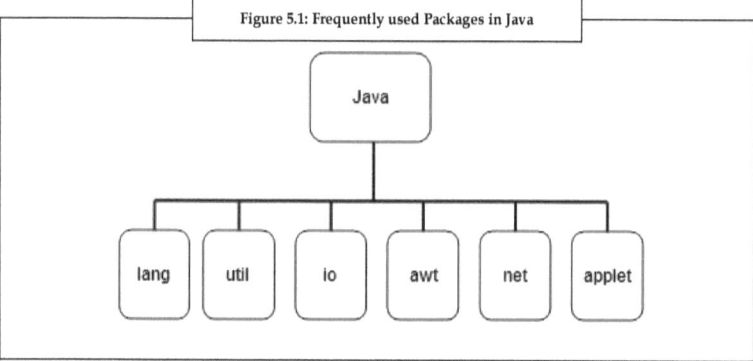

Figure 5.1: Frequently used Packages in Java

Table 5.2 illustrates the functionality of these frequently used packages.

Table 5.2: Functionality of Frequently used Packages

Packages	Functions
java.lang	Contains language support classes. These packages are automatically imported as these classes are used by the Java compiler itself. The **java.lang** package includes classes for primitive types, strings, math, functions, exception and threads.
java.util	Contains the language utility class such as vectors, hash tables, random numbers, date, and so on.
java.io	Contains the input/output support classes. These packages provide facilities for the input/output of data.

Cont..

java.awt	Contains a set of classes that are used for implementing graphical user interface. This package includes classes for buttons, windows, menus, lists, and so on.
java.net	Contains the classes that are required for networking. This package includes classes that are required for communicating with the local computers and also with the Internet servers.
java.applet	Contains the classes that are required to create and implement applets.

The two most widely used Java API packages are **java.lang** and **java.util**.

java.lang

As depicted in table 5.3, **java.lang** package is automatically imported by the Java compiler. It contains all the classes that are the basis for almost all the programs written in Java.

Table 5.3 shows the classes that are included in **java.lang** package.

Table 5.3: Classes Included in java.lang Package

Boolean	Byte	Character	Class	System
ClassLoader	Compiler	Double	Float	Thread
InheritableThreadLocal	Integer	Long	Math	ThreadGroup
Number	Object	Package	Process	ThreadLocal
Runtime	RuntimePermission	SecurityManager	Short	Throwable
StackTraceElement	StrictMath	String	StringBuffer	Void

The following are the interfaces that are defined by **java.lang** package:

1. Cloneable

2. Comparable

3. Runnable

4. CharSequence

The **Comparable** interface was added by Java 2 and the **CharSequence** interface was added by Java 2, version 1.4.

Notes

Most of the classes that are included in the java.lang package are deprecated methods. These methods were included to be implemented in Java 1.0, but are still provided by Java 2 to support the pool of legacy code. However, these deprecated methods are not used in new codes.

Java 2 added quite a lot of new classes and methods to the **java.lang** package.

To know the list of **java.lang** interfaces, classes and their functionalities, refer "Schildt, H. (2008). The Complete Reference, 7th ed. Tata McGraw-Hill."

java.util

One of the most powerful subsystems of Java is **collections**. The **java.util** package contains this subsystem – **collections**. A collection is nothing, but a group of objects. Collections were introduced in Java 2 and later enhanced by Java 2, version 1.4.

Apart from collections, **java.util** contains a variety of classes and interfaces that support various functionalities. These classes and interfaces are mostly used throughout the Java packages and in the Java codes written by programmers. Since **java.util** has several features, it is one of the most widely used Java packages.

Table 5.4 shows all the classes that are included in **java.util** package.

Table 5.4: Classes Included in java.util Package

AbstractCollection	Calendar	HashMap	Locale
AbstractList	Collections	HashSet	Observable
AbstractMap	Currency	HashTable	Properties
AbstractSequentialList	Date	IdentityHashMap	PropertyPermission
AbstractSet	Dictionary	LinkedHashMap	PropertyResourceBundle
ArrayList	EventListenetProxy	LinkedHashSet	Random
Arrays	EventObject	LinkedList	ResourceBundle
BitSet	GregorianCalendar	ListResourceBundle	SimpleTimeZone
Stack	StringTokenizer	Timer	TimerTask
TimeZone	TreeMap	TreeSet	Vector
WeakHashMap			

Just like the **java.lang** package, **java.util** package also defined few interfaces. These interfaces are depicted in the table 5.5.

Table 5.5: Interfaces Included in java.util Package

Collection	Comparator	Enumeration	EventListener	Iterator
List	ListIterator	Map	Map.Entry	Observer
RandomAccess	Set	SortedMap	SortedSet	

The ResourceBundle, ListResourceBundle, and PropertyResourceBundle classes are useful in the internationalization of large programs with many locale-specific resources. PropertyPermission allows you to grant read/write permission to a system property.

Notes   Most of the java.util interfaces were added by Java 2.

To know the list of **java.util** interfaces, classes and their functionalities, refer "Schildt, H. (2008). The Complete Reference, 7th ed. Tata McGraw-Hill."

Lab Exercise   Write a program to create and import a user defined package.

5.6 Summary

- Packages are similar to header files that are used in C and C++.
- Packages are mainly used for avoiding class conflicts and for facilitating the code reusability.
- A package can be created by using the package command as the first statement in a Java source file.
- Hierarchy of packages can be created using a period in between the names of packages.
- Packages can be classified as predefined packages or user-defined packages. Predefined packages are packages that are already defined. User-defined packages are packages that are created by users.
- Packages are identified with their respective directories.
- Access protection is done to protect variables from external modification.
- A package can be imported by using the fully qualified name of the package, importing the package and the class, or importing the package completely.
- Java API facilitates the grouping of several packages into different packages according to their functionality.
- The two most widely used Java API packages are java.lang and java.util.

5.7 Keywords

API: A set of rules and specifications that a software program can follow to access and use the services and resources provided by another particular software program that implements the same API.

Case-sensitive: Computer command that specifies that every letter in a word must be typed exactly as required, that is, in upper-case or lower-case.

Deprecate: Software features that are outdated and must be avoided.

Domain Name: Represents the symbolic depiction of numerical Internet address.

5.8 Self Assessment

1. State whether the following statements are true or false:
   (a) Packages are often stored in a corresponding folder in the file system, but packages can be stored in a database.
   (b) Reusability of code is possible only in inheritance.

(c) Java run-time system always uses the current working directory as its starting point.

2. Fill in the blanks:

   (a) _____defined within a package cannot be accessed by code outside that package.

   (b) While naming a package, first the _____has to be specified.

   (c) The user can also specify a directory path or paths by setting the _____environmental variable.

3. Select a suitable choice for every question:

   (a) Packages are mainly used to:

      (i) Facilitate reusability of code.

      (ii) Create hierarchy of packages.

      (iii) Manage library files.

      (iv) Specify a directory path.

   (b) Which of the following are the elements that a package encapsulates?

      (i) Data and code

      (ii) Variables and methods

      (iii) Classes and interfaces

      (iv) Classes and objects.

   (c) Identify which of the following packages are not declared as protected.

      (i) Subclasses that are in the same package.

      (ii) Subclasses that are in different packages.

      (iii) Non-subclasses that are in the same package.

      (iv) Non-subclasses that are in different packages.

   (d) Identify which of the following classes are included in java.util.

      (i) SecurityManager

      (ii) Throwable

      (iii) StackTraceElement

      (iv) Dictionary

   (e) Which of the following are useful in the internationalization of large programs with many locale-specific resources?

      (i) SortedSet

      (ii) RandomAccess

      (iii) ListIterator

      (iv) ListResourceBundle

5.9 Review Questions

1. "In Java, programmers can define their own packages, if there are similar classes and the classes are needed in more than one application". Explain how this can be achieved.

2. "Java run-time system knows where to look for a package that was created by a user and is used in a program". Comment.

3. "Packages add a different dimension to access control". Elaborate.

4.   "To use a package in any class, the package has to be imported in the class which has to use it". Discuss.

5.   "Java Application Programming Interface (API) facilitates the grouping of a large number of classes into different packages". Do you agree? Justify your answer.

6.   "The java.lang package is automatically imported by the Java compiler". What are the classes and interfaces that are contained in this package?

7.   "Packages are the library files in java". Justify.

8.   "Creating a package is simple". Explain the method that supports this statement.

9.   Explain the different packages that are supported by Java and also the methods of creating them.

10. "Adding a class to an already existing package is done by adding the statement with the classname.java source file". Elaborate.

Answers: Self Assessment

1.   (a) True

     (b) False

     (c) True

2.   (a) Classes

     (b) Domain name

     (c) CLASSPATH

3.   (a) Facilitate reusability of code

     (b) Variables and methods

     (c) Non-subclasses that are in different packages

     (d) Dictionary

     (e) ListResourceBundle

5.10 Further Readings

Books

Balagurusamy, E, Programming with Java 3e Primer, Tata McGraw Publishers, New Delhi

Schildt Herbert, Java A Beginner's Guide, Third Edition

Online link

http://java.sun.com/docs/books/jls/third_edition/html/packages.html

http://download.oracle.com/javase/1.4.2/docs/api/java/util/package-summary.html

Unit 6: Interfaces

Objectives

After studying this unit, you will be able to:

• Define interfaces

• Describe the implementation of interfaces

• Explain the use of extends keyword to extend interfaces

Introduction

In software engineering, we come across several situations where groups of programmers find it necessary to come to a common 'agreement' that indicates how their software interacts. Each group must be in a position to develop a code without having any knowledge on the code that is being developed by the other group. In Java, **interfaces** are such 'agreements'. The concept of interface was introduced to achieve multiple inheritance, hierarchical inheritance, and dynamic polymorphism.

An interface is a unique code in Java, which contains method signatures (includes the method name and the parameter list), and some constant values. Interfaces are like abstract classes, but, the major difference between inheritance and abstract classes is that, in interfaces, all the methods are abstract, whereas abstract class can also contain methods that have its implementation in the same class other than the abstract methods. Interface looks like a class, but without instance variables, and its methods are declared without any coding in the body of the method. The implementations of these methods are given in the class that implements this interface.

Interfaces specify the structure of the class by declaring the methods with their signatures. The keyword **implements** is used in the class, while declaring the class name, that is, at the time of class definition. When a class implements an interface, it must give the definition for all the methods that are defined in the interface. If any method of the interface is not defined in the class, then the compiler gives an error message while compiling the class; in such cases, the class must be declared as an **abstract** class.

A single interface can be implemented in more than one class. Similarly, a class can implement more than one interface. Objects or instances of an interface cannot be created. Moreover, interfaces do not have their own behavior. The methods of an interface are always public, even if not declared public. The constant variables that are declared in the interface are always **final**, even if not declared final.

6.1 Defining Interfaces

An interface is defined just the same way as a class is defined.

General Form of Defining an Interface

    access specifier interface [interface name]
        {
            // Body of the interface
                type final var a = value;
                type final var b = value;
                return type method nameA(Parameter list);
                return type method nameB(Parameter list);
        }

In this syntax, an **access specifier** such as **public**, **private** or **protected** must be used. If the access specifier is not declared, then the interface takes the default access specifier (public), and is available to the other classes or interfaces, which are only within the same package. After the access specifier, the keyword **interface** is given followed by the name of the interface. The body of the interface is written within the braces, and it consists of the constant variables and the method declaration.

> *Example:* public interface Int1
>   {
>   public void exam( );
>   }
>
> In this example of a simple interface, the interface **Int 1** has only one method, that is, **public void exam( );**. This method is implemented in the class of classes that implement the interface **Int 1.**

It is important to define an interface, so that it can be used in more than one class that has similar methods.

6.2 Implementing Interfaces

Once defined, interfaces are implemented. Implementing interfaces helps to access the methods and variables of the interface in a class. When a class implements an interface, all the methods in the interface have to be implemented in that class. If there is any method that has not been implemented, then the Java compiler gives an error message while compiling the program.

Once the interfaces are defined, they can be implemented by one or more classes. For implementing an interface in any class, a keyword **implements** is used in the definition of the class. The keyword **implements** indicates that the methods that are defined in the interface are implemented in the class. The class that implements the interface must be declared as **public,** and it is free to have its own variables and methods.

General Form of Implementing an Interface

    access modifiers class [class name] implements [Interface name]
        {
            Type final var = value;
            Return type method name(Parameter list);
        }

The above syntax is of a class implementing an interface. An **access modifier** must be specified in the syntax, along with the **class name**, followed by the keyword **implements** and the **name of the interface** that has to be implemented. Within the braces, the methods and constant variables are declared.

 Example: Program to illustrate the implementation of an interface.

```
public class Interfaceexample implements Int1
{
public void exam( )
{
System.out.println("Example of an interface");
}
public static void main(String args[ ])
{
Interfaceexample intex = new Interfaceexample( );
intex.exam( );
}
}
```

Output:

Example of an interface

In this example,

1. First, a class **Interfaceexample** implements the interface **Int1**.
2. Then, the method **exam( )** of the interface **Int1** is implemented in the class **Interfaceexample.**
3. The **main( )** method of the class is then called.
4. In this **main( )** method,
    (a) A new object intex of the class **Interfaceexample** is created, by using the **new** keyword.
    (b) Then, **exam( )** method of the object **intex** is called.

Finally, the program is tested and executed. The output of the application prints the implementation given in the class **Interfaceexample** for the method of the interface **Int1.**

Notes A class can implement more than one interface.

Task Write a program to implement an interface in the class.

6.3   Extending Interfaces

An interface can extend another interface just like how a class can extend another class. As an **extend** keyword is used to extend classes, interfaces also use **extends** keyword to extend interfaces.

General Form of Extending an Interface

access modifier interface [interface name] extends interface 1, interface 2, ...

    {

        Type final var = value;

        Return type method name(parameter list);

    }

In the above syntax, first the **access modifier** is specified along with the keyword **interface** and the name of the interface to which the other interfaces are inherited. This is followed by the keyword **extends** and the name of interfaces which are being inherited. Within the braces, the methods and constant variables are declared.

Example: public interface One

{

public void extend1( );

}

The above code is an example of an interface, which has a single method extend1( ); .

public interface Two extends One

{

public void extend2( );

}

The above code is an example of an interface, which extends another interface. Here, **interface Two** is the interface, which inherits **interface One**. Hence, the keyword extends is used in the definition of the interface.

Combining the previous two examples:

Example:

```
public class Intextend implements Two
{
public void extend1( )
{
System.out.println("Implementing from interface One");
}

public void extend2( )
{
System.out.println("Implementing from interface Two");
}
public static void main(String args[ ])
{
Intextend ie = new Intextend( );
ie.extend1( );
ie.extend2( );
}
}
```

Output:
Implementing from interface One
Implementing from interface Two

In this example,
1. First, a class **Intextend** implements another class **Two.**
2. In this class,
    (a) The **extend1( )** method is called. In this method,
        System.out.println("Implementing form interface One"); statement is
        used to print Implementing form interface One on the screen.
    (b) Then, the extend2( ) method is called. In this method,
        System.out.println("Implementing form interface Two"); statement is
        used to print Implementing form interface Two on the screen.
    (c) The **main( )** method of the class is then called.

(d) In this **main( )** method,

   (i) A new object **ie** of the class **Intextend** is created, by using the **new** keyword.

   (ii) Then, the methods **extend1( )** and **extend2( )** are called for the new object.

Finally, the output of the application prints the implementation given in the class **Intextend** for the method of the interfaces **One** and **Two.**

Notes Classes cannot extend interfaces and interfaces cannot extend classes.

Task Write a program to extend interfaces and implement it in the class.

Accessing Interface Variables

The variables of an interface are always declared as **final.** Final variables are those variables, whose values are constant and cannot be changed. The class that implements the interface can use the variables as declared in the interface and cannot modify or change the value of the variable.

  Example: Program to illustrate the concept of accessing interface variables.

```
public interface Selectcolor
{
    int blue = 4;
    int yellow = 5;
    int pink = 6;
    public void choose(int color);
}
```

In the above code, an interface named as Selectcolor is given. The interface has three variables; all are integer variables. It has a method void choose, which takes an integer type parameter.

```
class SelectImp implements Selectcolor
{
    public void choose(int color)
    {
        switch(color)
        {
        case blue: System.out.println("The color selected is blue");
```

```
            break;
        case yellow: System.out.println("The color selected is yellow");
            break;
        case pink: System.out.println("The color selected is brown");
            break;
        }
    }
    public static void main(String args[ ])
    {
        SelectImp si = new SelectImp( );
        si.choose(4);
        si.choose(5);
        si.choose(6);
    }
}
```

Output:

The color selected is blue

The color selected is yellow

The color selected is brown

In the above example,

1. First an interface **Selectcolor** is created and the values for the integers blue, yellow and brown are set as 4, 5 and 6 respectively.

2. Then a method **choose( )** which takes in an integer parameter is declared.

3. A class **SelectImp** is created, which implements the interface **Selectcolor.**

4. Then the method **choose( )** of the interface **Selectcolor** is implemented using the switch case statements.

5. Then there is a **main( )** method which creates the object of the class **SelectImp** and call the **choose( )** method of the **SelectImp** class with different parameters or argument.

    Then the program is tested and executed.

If the integer value that is passed as an argument to the **choose( )** method is '4' which is the value for the color 'blue' then the first case is executed and the statement "The color selected is blue" is printed. If the value passed as an argument to the choose( ) method is '5' then the second case statement is executed and thus the output will be "The color selected is yellow," and so on.

In this application, the choose( ) method is called three times in the main( ) method with different values so all the three cases are executed.

The output of the application prints the implementation given in the class **SelectImp** for the method of the interface **Selectcolor**.

6.4 Differences between Packages and Interfaces

Table 6.1 shows the differences between packages and interfaces.

Table 6.1: Differences between Packages and Interfaces

Packages	Interfaces
A package is a method used to group objects; it is very similar to grouping items within a folder or directory on a file system.	An interface is a .java file that is implemented by another class to tell the outside world that it corresponds to a certain specification.
A class is present within a package, but this does not have an impact on the behavior of the class.	Interfaces basically refer to the visible methods of an object.

Interfaces are more similar to abstract classes than packages. An interface does not contain any implemented methods. On the other hand, an abstract class can define few methods and allow few methods to be implemented by a subclass. Another difference between interfaces and abstract class is that, a class can implement multiple interfaces, but a class can extend only one abstract class.

Lab Exercise   Write a program to access an interface variable.

6.5 Summary

- Interfaces are unique codes that contain method signatures and few constant values.

- It is possible to implement a single interface in multiple classes. A class can also implement multiple interfaces.

- The methods of an interface are always public.

- An interface can extend another interface.

- Extends keyword is used to extend interfaces.

- Interfaces are more similar to abstract classes.

6.6 Keywords

Access Specifier: Controls the access to the names that follow it, up to the next access specifier or the end of the class declaration.

Dynamic Polymorphism: Polymorphism exhibited at runtime.

Hierarchical Inheritance: Multiple classes being derived from a single class.

Polymorphism: Ability of an object to take on many forms.

6.7 Self Assessment

1. State whether the following statements are true or false:
   (a) Implementing interfaces helps to access the methods and variables of the interface in the class.
   (b) Cloneable is an interface that is included in java.util.
   (c) The class that implements the interface can use the variables as declared in the interface and can modify or change the value of the variable.

2. Fill in the blanks:
   (a) The variables of an interface are always declared as _____.
   (b) An _____is a java file that is implemented by another class to tell the outside world that it corresponds to a certain specification.
   (c) Interfaces use _____keyword to extend interfaces.

3. Select a suitable choice for every question:
   (a) Which of the following keyword is used in the class while declaring the class name?
      (i) Implements
      (ii)
      Extends(iii)
      Abstract   (iv)
      Final
   (b) The class that implements the interface must be declared as _____.
      (i) private
      (ii) public
      (iii) protected
      (iv) default

6.8 Review Questions

1. "Interfaces are like abstract classes". Comment.
2. "Once the interfaces are defined, it can be implemented by one or more classes". How can this be achieved?
3. "An interface can extend another interface just like how a class can extend another class". Explain the methods to achieve this.
4. "Interfaces are similar to abstract classes than packages". Discuss.
5. "An interface is defined just the same way as class is defined". Justify.
6. "The class that implements the interface can use the variables as declared in the interface and cannot modify or change the value of the variable". Discuss.

Answers: Self Assessment

1. (a) True   (b) False   (c) False
2. (a) final   (b) Interface   (c) extends
3. (a) Implements   (b) public

6.9 Further Readings

Books

Balagurusamy, E, Programming with Java 3e Primer, Tata McGraw Publishers, New Delhi

Schildt Herbert, Java A Beginner's Guide, Third Edition

Online link

http://www.learn-java-tutorial.com/Java-Interfaces.cfm

http://download.oracle.com/javase/tutorial/java/IandI/createinterface.html

Unit 7: Exception Handling

Objectives

After studying this unit, you will be able to:

- Explain the fundamentals of exception
- Describe uncaught exception
- Discuss throw, throws and finally keywords
- Explain the creation of exception subclasses

Introduction

A program rarely runs successfully in its first attempt. This is because of the errors that are present in the program code. In many programming languages, the errors that occur in programs must be identified and corrected manually. This is a troublesome process. But, Java has such a mechanism of exception handling that helps the programmers to catch the exceptions easily and handle these exceptions carefully, so that it does not affect the programs during execution.

According to Bruce Eckel, "The reason exception handling systems were developed is because the approach of dealing with each possible error condition produced by each function call was too onerous, and programmers simply weren't doing it. As a result, they were ignoring the errors. It is worth observing that the issue of programmer convenience in handling errors was a prime motivation for exceptions in the first place".

Exception handling mechanism has one main advantage, that is, much of the error handling code that previously had to be entered manually, can now be automated into any large program.

 Example: In some programming languages, error codes are returned on the failure of a method, and these error codes must be checked manually, whenever the method
is called. This methodology is both difficult and error-prone.

7.1 Meaning of Exception Handling

The problem that comes up during the execution of a program is denoted as an exception. This problem can arise due to several reasons, which includes:

1. Supply of invalid data by user

2. Unavailability of the file required to be accessed

3. Loss of network connection in the middle of communications

4. Shortage of memory in the JVM

Some exceptions occur due to the error generated by the user and the programmer, whereas some other exceptions occur due to the failure of physical resources.

An exception can be considered as an object that includes information about the type of error that has occurred.

Whenever any problem occurs while accessing a disk file, Java generates an exception in the form of an object specifying the details of the problem. Thus, whenever a network related problem is identified, Java wraps up the details of the problem into an object in the form of an exception.

Whenever an error occurs, an object that represents the exception created and is thrown to the method in which the error occurred. The exception that occurred has to be handled in the same method or can be passed on further in the application, but it has to be caught in the application at some point and processed, so that the application does not give any error during run time. The exception object can be generated by the run time system or can be created manually. When an exception object is created and handed over to the run time system, it is called as **throwing an exception**.

Apart from generating exceptions for the identification of errors that are external to a program, Java also generates exceptions for irregular code within the program.

    Example: Java produces an exception whenever an invalid array of index is accessed or an illegal class-cast is attempted.

Exception handling follows the concept of universal error processing, wherein the error correction code is taken out from the main body of code and is supplied to several exception handlers.

It can also be said that exceptions are necessary to interrupt the flow of control, when something important or unpredicted (generally an error) occurs. Basically, whenever an exception is raised, the control is transferred to some other part of the program that can try to deal with that exception/error, or at least terminate it completely.

Once an exception occurs, the program is automatically terminated. This exception must be handled properly to run that program. Exception handlers are used to handle these exceptions. Java uses the following keywords to handle exceptions:

1. try

2. catch

3. finally

4. throws

5. throw

When only **try** and **catch** are used in a program, their combination is called as the **try-catch** block. This block is used to catch a Java exception. Only one exception type can be handled by every **catch** block, and more than one **catch** clause can be used in a single **try** block. In the **try-catch** block, the **try** block surrounds a statement that may cause the occurrence of the exception, and the **catch** block follows the **try** block. On the occurrence of the exception, a code (that should be executed) is specified in this **catch** block.

Some of the terminologies used within exception handling are:

1. *Throwing:* A process through which an exception is generated and thrown to the program.

2. *Catching:* Capturing a currently occurred exception and also executing statements that may help in resolving those exceptions.

3. *Catch Clause, or Catch Block:* The block of code that tries to handle the exception.

4. *Stack Trace:* A series of method calls that brings back the control to the point where the exception had occurred.

An exception handler comprises two core sections, namely:

1. *The Try Block:* It includes some code, which might throw an exception (generate an error).

2. *The Catch Block:* It includes the error handling code. It means that it determines the strategy to be implemented, when an error is detected.

Thus, exception handling provides a method to separate error handling from the code that may result in errors. It is beneficial in several cases, as it produces **clean executable code**.

Notes **Factors to be Considered for Exception Handling**

1. Exception handling does not reduce the amount of work required while handling errors of small programs.

2. The main advantages of exception handling can be observed in a program of considerable size.

When to Use Exception Handling

Exception handling can be implemented to process exceptions occurring at many points, such as:

1. For processing exceptional situations and errors that cannot be expected logically, that is, situations, which are rarely expected to occur.

2. For processing exceptions from libraries and classes that cannot be expected to handle exceptions on their own like, network connections.

3. For managing error handling in a consistent manner throughout the project, in case of large projects. This could be achieved by using global exception handlers.

Local Exception Handlers: The **catch** blocks that are employed immediately after the **try** block in which the exception was thrown are referred to as local exception handlers.

Global Exception Handlers: The **catch** blocks employed at the higher level of the call hierarchy, projected to achieve universal error handling are referred to as global exception handlers.

When Not to Use Exception Handling

1. Exception handling must not be used in situations, where it is not much applicable. Moreover, we can only implement exception handling for exceptional events, that is, for errors that rarely occur.

 Example: After requesting the user to supply input value between 6 and 12, it would not be appropriate to employ exception handlers to process any errors that are out-of-range; rather a more traditional error checking technique should be used.

2. Similarly, we must not employ exception handlers to handle errors related to key presses, mouse clicks, network messages, and so on, since these are better restricted to event or interrupt processing.

7.1.1 Exception Types

Classification of exceptions in Java is based on the way the Java compiler handles it. In Java, exceptions are categorized into two types, as shown in figure 7.1.

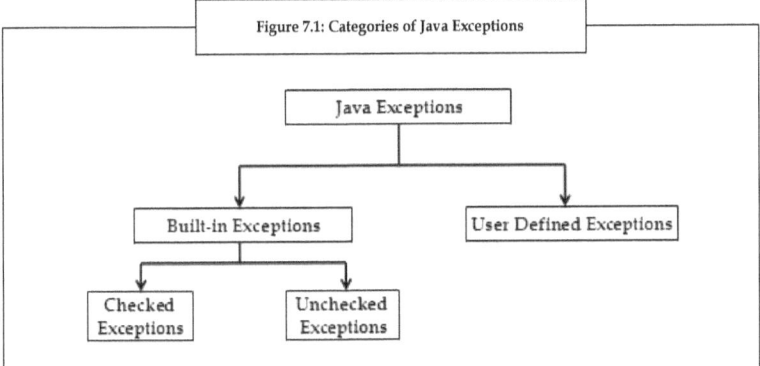

As shown in the figure 7.1, there are two categories of Java exceptions, namely, built-in exceptions, and user-defined exceptions, where, built-in exceptions are further classified into checked exceptions and unchecked exceptions.

Built-in Exceptions

As the name suggests, built-in exceptions are those exceptions that are already built-in in a programming language (like Java). Java runtime system throws such exceptions for describing the condition of exceptions. In Java, there are two types of built-in exceptions, which are checked exceptions and unchecked exceptions.

Checked Exceptions

Checked exceptions are those built-in exceptions that occur at the time of compilation of a program. At the time of compilation, the compiler verifies if the checked exception handlers are contained in the program or not. These exceptions are not inherited from RuntimeException class, but from java.lang.Exception class. These exceptions must be handled properly by the programmer, for avoiding any compile-time error. Examples of checked exceptions are:

1. ***NoSuchFieldException:*** This exception is thrown when the user tries to use any field or variable in a class that does not exist.

2. ***ClassNotFoundException:*** This exception is thrown when the user tries to access a class that is not defined in the program.

3. ***IllegalAccessException:*** This exception is thrown when the access to a class is denied.

4. ***Interrupted Exception:*** This exception is thrown when a thread is interrupted in processing, waiting, or sleeping state.

5. ***NoSuchMethodException:*** This exception is thrown when the user tries to access a method that does not exist in the program.

Unchecked Exceptions

Unchecked exceptions, also called as runtime exceptions, are those built-in exceptions that occur during the program runtime, and are internal to the application. Such exceptions are derived from the java.lang.RuntimeException, which is inherited from the java.lang.Exception class. The user cannot predict and recover from these exceptions. Usual causes of such exceptions are data errors like arithmetic overflow, division by zero, and so on.

Examples of unchecked exceptions are:

1. **StringIndexOutOfBoundsException:** This exception is thrown when a program tries to access a character at an index of the string that does not exist.

2. **ArrayIndexOutOfBoundsException:** This exception is thrown when the user tries to access an array index that does not exist.

3. **ArithmeticException:** This exception is thrown when the user tries to divide a number by zero.

4. **NullPointerException:** This exception is thrown when the Java Virtual Machine (JVM) tries to execute some operation on an object which points to a null data or no data.

User-defined Exceptions

Just like built-in exceptions, user-defined exceptions also exist in Java. As the name suggests, user-defined exceptions are those exceptions that are defined by the user of the program. For creating an exception type, the user has to extend the Exception class and create their own subclass exception class. To do so, the user has to inherit the Exception class. Exception classes follow a hierarchy, where Throwable class is the parent class for an entire family of exception classes, which is declared in **java.lang** package as **java.lang.Throwable**. This **throwable** class can be divided into two types, namely, exceptions, and errors (exceptions defined by the user).

Caution   The exception class defined by the user must be a subclass of Exception class.

Caution   The toString( ) method must be superseded in the user-defined exception class, for displaying meaningful information about the exception.

Handling and Creating User-defined Exception

Keywords such as **try**, **catch** and **finally** are used in implementing user-defined exceptions. This Exception class inherits all the methods from **Throwable** class. These methods are depicted in the table 7.1.

Table 7.1: Methods in Throwable Class

Methods	Explanation
String toString( )	Provides description of the exception and returns a String object.
String getMessage( )	Describes the exception in program.
Throwable fillInStackTrace( )	Returns a **Throwable** object that includes a stack trace.
void print StackTrace( )	Returns the stack trace.
void printStackTrace(PrintStream stream)	Returns the stack trace to the stream defined.
String getLocalizedMessage	Returns the localized definition of the exception.

Example: Program to create a user defined exception.

```
class UDefExcept extends Exception
{
        String note = "";
        int marks;
        public UDefExcept ( )

        {

        }
        public UDefExcept (String s1)

        {
          super(s1);
        }
        public String toString( )

        {
          if (marks <= 40)
          note = "You need to improve";

          if (marks > 40)

          note = "You are the winner";
          return note;
        }
}
public class test
{
    public static void main(String args[ ])

    {
      test t = new test( );
      t.tm( );
    }
    public void tm( )

    {
      try

      {
        int j=0;

        if( j < 40)
        throw new MyExcept( );

      }
      catch(MyExcept ee1)

      {
        System.out.println("my ex"+ee1);
      }
    }
}
```

Output:

On the command prompt –

C:\ >javac test.java

C:\ >java test
my ex You need to improve

In this example,

1. First, a class **UDefExcept** is created from the **Exception** class using the **extends** keyword.

2. In the class **UDefExcept**,

   (a) A String **note** is declared with no value assigned to it.

   (b) Then, an integer **marks** is declared.

   (c) The **UDefExcept( )** constructor is then called without any parameters, and no method within.

   (d) The **UDefExcept (String s1)** constructor is again called, but with a **String s1**. In this constructor, the **super( )** method is called on the **String s1**.

   (e) Then, **String toString( )** method is called to provide description of the exception and return a String object. In this method,

      (i) The **if** statement is used to check the **marks <= 40** condition. If this condition is true, **note** is assigned a string **You need to improve**.

      (ii) Another if statement is used to check the marks > 40 condition. If this condition is true, **note** is assigned a string **You are a winner**, and this string is returned using the **return** statement.

3. Another class test is created.

4. In the class test,

   (a) The main( ) method of the class is called.

   (b) In the main( ) method,

      (i) A new object t of the class test is created using the new keyword.

      (ii) Then, t.tm( ); statement is used to call the method tm( ) on the object t.

   (c) Then, the tm( ) constructor is called and declared void.

   (d) In this constructor, try-catch mechanism is used, wherein

      (i) In the try block, first, an integer j is declared and assigned the value of 0. Then, the if statement is used to check the condition j<40. If this condition is true, a new exception MyExcept( ); using the throw and new keywords is thrown.

      (ii) Then, the catch block is used to catch the exception MyExcept ee1. In the catch block, the System.out.println("my ex"+ee1); statement is used to print my ex and the value of ee1 is printed on the screen.

In this example, the exception is thrown when the value of **marks** is less than 50.

7.2 Uncaught Exception

The main logic that we have been discussing so far applies to checked exceptions, or exceptions, whose flow is **explicitly controlled**. As per this logic, a method can throw a given exception, and the caller to that method must explicitly handle the exception or **throw** it to the next caller.

Occurrence of an Unchecked Exception

An uncaught exception that is also referred to as an unchecked exception, cannot be caught in a **try** or **catch** block. Whenever an unchecked exception occurs (such as, in a simple single-threaded program), Java prints out the stack trace of the exception and then terminates the program.

 Example: Program to illustrate what happens when an unchecked exception occurs.

```
public class TestExceptions
{
public static void main(String args[ ])
{
String str = null;
int length = str.length( );
}
}
```

Output:

Exception in thread "main" java.lang.NullPointerException

at test.TestExceptions.main(TestExceptions.java:4)

In this example,

1. First, a class **TestExceptions** is created and is declared public, by using the **public** keyword.

2. In this class TestExceptions,

   (a) The main( ) method of the class is called. In this main( ) method,

   (b) A string str is assigned a null value.

      (i) A method str.length( ); is called to check the length of the string, and this value is assigned to integer length.

      (ii) An integer int

Note: This example shows the NullPointerException.

Handling Uncaught Exceptions

Unchecked exceptions are not caught in a try/catch block. It can often be observed that Java prints the exception stack trace and then terminate the program. But, actually this is a general view of what happens.

Whenever an uncaught exception originates in a particular thread, Java looks for what is called an **uncaught exception handler**, which is mainly an implementation of the interface **UncaughtExceptionHandler**. This interface has a method **handleException( )**, which the implementer dominates to take suitable action, such as printing the stack trace to the console. It is also possible to install your own instance of **UncaughtExceptionHandler** to deal with the uncaught exceptions of a particular thread, or even for the whole system.

Did you know?   Java mainly deals with uncaught exceptions according to the thread in which they arise.

Whenever an uncaught exception arises, the JVM (Java Virtual Machine) does the following tasks:

1. A special private method **dispatchUncaughtException( )** is called on the Thread class, in which the exception occurs, and then the thread in which the exception occurred is terminated.

2. In turn, the **dispatchUncaughtException( )** method calls the thread's **getUncaughtExceptionHandler( )** method to search for a proper uncaught exception handler to use. Generally, this would be the thread's parent **ThreadGroup**, whose **handleException( )** method will print the stack trace by default.

Though, this process can be overridden for a particular thread, for a **ThreadGroup**, or for all threads, as shown in the table 7.2.

Table 7.2: Thread Handlers and Implementation		
Thread Handler to Set	**How to Set**	**Notes**
All threads	Thread.setDefaultUncaughtExceptionHandler ( )	Relies on a ThreadGroup's **uncaughtException( )** method not being overridden, or on any overriding implementation searching for the default handler.
All threads of a particular thread group	Override ThreadGroup.uncaughtException( )	Means a **ThreadGroup** subclass has to be provided.
Particular thread	Thread.setUncaughtExceptionHandler( )	You can also override **getUncaughtExceptionHandler( )**, if using your own **Thread** subclass.

The process that determines which uncaught exception handler to call is shown in the figure 7.2.

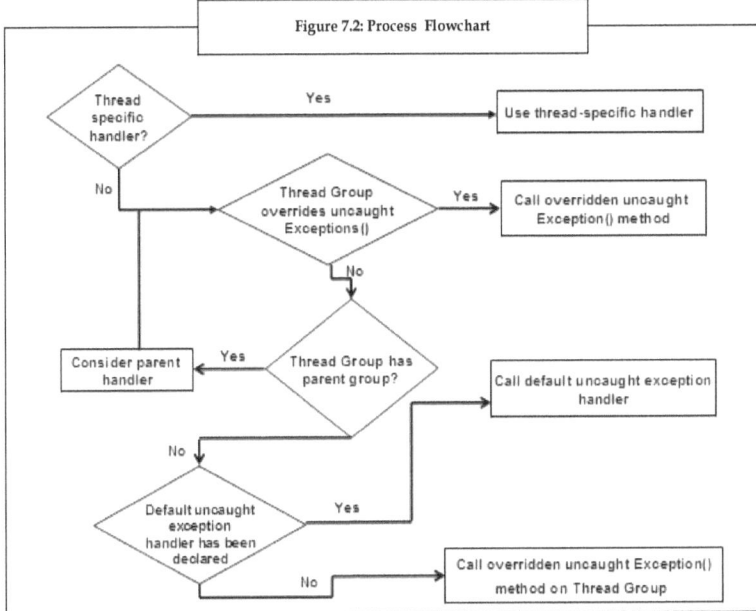

Figure 7.2: Process Flowchart

The figure 7.2 describes a process which decides on which uncaught exception handler to call for a given thread.

Notes

Care should be taken to stop uncaught exceptions from terminating essential threads or compensating for thread termination.

Preference Given to Uncaught Exceptions

Uncaught exceptions behave differently from other exceptions. They are preferred for the following reasons:

1. They need not be caught explicitly. Whenever an uncaught exception, such as NullPointerException, ClassCastException, OutOfMemoryError occurs, Java handles them automatically.

2. Methods and constructors need not be specified explicitly as they can throw an unchecked exception since it is taken for granted that any method or constructor can throw them.

3. Moreover, certain Java byte-code instructions such as accessing an array, invoking a method on an object, division of integers, can also throw an unchecked exception.

Caution

Do not suppress or ignore exceptions.

7.3 Throw

In Java programs, an exception is forced by using the **throw** keyword. This keyword can also be used for passing a custom message to the exception handling module. It is possible for any program to throw an exception explicitly, using the keyword 'throw'.

General Syntax of Throw

throw ThrowableInstance;

In this syntax, **throw** is the keyword, and the **ThrowableInstance** must be a **Throwable** object or a **Throwable** subclass. Some simple types, such as **int** or **char**, as well as **non-Throwable** classes, such as **string** and **object**, cannot be employed as exceptions. A **Throwable** object can be obtained by two ways, either by using a parameter into a **catch clause**, or by creating one with the **new** operator.

Once the **throw** statement is encountered, the flow of execution stops immediately and none of the consequent statements are executed. The closest enclosing try block is examined to find out if it has a catch statement that matches the type of the exception. If the match is found, control is transferred to that statement. If no match is found, the next enclosing **try** statement is examined, and so on. If no matching catch is obtained, the default exception handler stops further processing of the program and prints the stack trace.

 *Example:* Program to illustrate the concept of creation and throwing an exception. The handler that catches the exception re-throws it to the outer handler.

```
// Demonstrate throw.
class ThrowDemo
{
    static void demoproc( )
    {
        try
        {
            throw new NullPointerException("demo");
        } catch(NullPointerException e)
        {
            System.out.println("Caught inside demoproc.");
            throw e; // rethrow the exception
        }
    }
    public static void main(String args[ ]) {
        try
        {
            demoproc( );
        } catch(NullPointerException e)
        {
            System.out.println("Recaught: " + e);
        }
    }
}
```

Output:

Caught inside demoproc.
Recaught: java.lang.NullPointerException: demo

This program deals with the same error twice.

1. First, a class **ThrowDemo** is created.

2. In the class **ThrowDemo**,

    (a) The demoproc( ) constructor is called, which is declared static and void.

    (b) In the demoproc( ) constructor,

        (i)   In the try block, throw new NullPointerException("demo");
              is used to throw a NullPointerException.

        (ii) Then, the **catch** block is used to catch the **NullPointerException** e.

        (iii) In the catch block, the Caught inside demoproc statement is printed
              on the screen. Also, the exception **e** is rethrown.

    (c) Then, the **main( )** method of the class is called.

    (d) In the **main( )** method,

        (i)   In the **try** block, **demoproc( )** method is called.

        (ii) Then, **catch** block is used to catch the **NullPointerException e**.

        (iii) In the **catch** block, the **Recaught:** statement with the value of
              exception **e** is printed on the screen.

Choosing the Type of Exception to be Thrown

When you have to choose among the type of exception to be thrown, you can choose anyone from the two provided below:

1. Use the one designed by someone else. The Java development environment offers a lot of exception classes that can be easily used.

2. Design one for your own use.

7.4 Throws

Apart from **try-catch** and **finally** clause, there is another way to handle an exception, which is, using the **throws** clause. Whenever a method, which throws a **checked** exception, is called from the Java API, the exception must be either thrown or caught. If it is not possible to handle the exception properly, then the declaration of the exception to the method header is done using the **throws** keyword, which is followed by the exception's class name.

In Java, the keyword **throws** signifies that the method raises a particular type of exception while being processed and it is applicable to that method only. The keyword **throws** in Java programming language takes a list of the objects as arguments of type **java.lang.Throwables** class.When the keyword **throws** is used along with a method, it is called as **ducking**. The method, which is calling another method along with a **throws** clause, is required to be enclosed within the try catch blocks.

Example: Sample code to illustrate the use of the **throws** keyword in a class.

```
import java.io.IOException;
public class Class1
{
  public method readingFile(String file) throws IOException
  {
    <statements>
    if (error)
    {
      throw new IOException("error reading file");
    }
  }
}
```

In this example,

1. First, the **java.io.IOException** package is imported.

2. Then, a class **Class1** is created.

3. In the class **Class1**,

    (a) The **readingFile(String file)** method is called, which throws **IOException**.

    (b) In this method,

         (i) First, the statements within that method are executed.

         (ii) Then, the if statement is used to check the error condition. If this error condition is true, a new IOException is thrown.

Note: This is not a complete program, rather a sample code from a program showing the use of **throws** keyword.

Notes **Factors to be Considered while Using Throws**

1. Sometimes, a method generates an exception that cannot be handled; the method must state this fact using a throws clause.

2. We do not need throws clause for any unchecked exceptions.

3. A throws clause is required only when the method is not able to handle the exception.

7.5 Finally

Finally blocks are executed once the **try** block exits. This block executes even after the unexpected exceptions has occurred. Irrespective of the **try** block, the expression in finally block is always executed. This block is in-built with the capability of recovering loss and preventing leak of resources. On closing and recovery of a file, the expression should be placed in the **finally** block.

The success of the **finally** block depends upon the existence of the **try** block and will execute only when an unexpected exception arises in the code.

Example:

```
import java.io.*;

public class FinallyException
{

    public static FileInputStream inputStream(String fileName)
        throws FileNotFoundException
    {
        FileInputStream fis = new FileInputStream(fileName);
        System.out.println("f1: File input stream created");
        return fis;
    }
    public static void main(String args[ ])
    {
        FileInputStream fis1 = null;
        String fileName = "saurabh.txt";
        try
        {
            fis1 = inputStream(fileName);
        } catch (FileNotFoundException ex)

        {
            System.out.println("FileNotFoundException occurred");
        } catch (Exception ex)

        {
            System.out.println("General exception occurred");
        }
        System.out.println( FinallyException.class.getName( ) + " ended");
    }
}
```

Output:

FileNotFoundException occurred
FinallyException ended

In this example,

1.   First, **java.io.*** package is imported.

2.   Then, a class **FinallyException** class is created.

3. In the class FinallyException,

   (a) The **FileInputStream inputStream(String fileName)** method is called with the string **fileName** in it, and is declared public and static.

   (b) The **FileNotFoundException** exception is thrown using the **throws** keyword. In this exception,

   (i) A new FileInputStream object fis is created.

   (ii) Then, the f1: File input stream created statement is printed on the screen.

   (iii) Also, the value of fis is returned using the return statement.

   (c) The **main( )** method of the class is called. In the **main( )** method,

   (i) The **FileInputStream fis1** is assigned the value **null**.

   (ii) The, **String fileName** is assigned the string **saurabh.txt**.

   (iii) In the **try** block, **fis1** is assigned the value in **inputStream**, that is, **filename**.

   (iv) Then, the **catch** block is used to catch the **FileNotFoundException ex** exception. In the **catch** block, the FileNotFoundException occurred statement is printed on the screen.

Second **catch** block is used to catch **Exception ex** exception. In this **catch** block, the **General exception occurred** statement is printed on the screen.

At last, the **getName( )** method is called to get the name of **FinallyException**, and prints it on the screen.

Task   Write a program to illustrate the use of **try** and **catch** and **finally** keywords.

7.6 Creating Exception Subclasses

Sometimes a plan is made to develop a package of Java classes working together. This plan is made to offer some useful functions to the users. A lot of efforts are put into it to ensure that the classes interact well with each other and that their interfaces are easy to recognize and use. Time is also spent on thinking about the design of the exceptions thrown by the classes.

Though most of the regular errors are processed by Java's built-in exceptions, the exception handling mechanism of Java is not restricted only to these errors. Generally, a fraction of the power of Java's approach to exceptions is its capability to handle exceptions created by the user, and which communicates with errors through the code specified. Creating an exception is very simple; just define a subclass of **Exception**, which is a subclass of **Throwable**. These subclasses do not require implementation of anything, since their existence in the type system permits them to be used as exceptions. The **Exception** class does not define any of its own methods, but it does inherit the methods provided by **Throwable**. Thus, the methods defined by **Throwable** are available for every exception, including those methods too, which are created by the user. It is also possible to override any of these methods in exception subclasses that you create.

Caution   Never use exceptions for flow control.

Naming Conventions

It is a good practice to attach the word **Exception** to the end of all classes that inherit either directly or indirectly from the Exception class. Likewise, classes that inherit from the Error class should be attached with the string **Error**.

Task

Create a user defined exception class that throws an exception string when compiled.

Lab Exercise

1. Write a program to divide a number with zero without raising any error.
2. Write a program to catch and process an error three times.

7.7 Summary

- An exception is an object that includes information about the type of error occurred.

- Exception handling is based on the concept of universal error processing, wherein the error correction code is separated out from the main body of code and is fed to several exception handlers.

- A checked exception is thrown whenever there is a probability of error in input-output processing.

- Unchecked exception is thrown due to the invalid argument supplied to a method. These exceptions originate at run-time.

- Keywords such as **try, catch,** and **finally** are used for implementing user-defined exceptions.

- An uncaught exception that is also referred to as an unchecked exception cannot be caught in a try or catch block.

- Once the **throw** statement is processed, the flow of execution stops immediately, and none of the consequent statements are executed.

- When the keyword **throws** is implemented along with a method, it is identified as ducking.

- The expressions in the **finally** block are always executed at run time.

- For creating an exception, just define a subclass of **Exception**, which is actually a subclass of **Throwable**.

7.8 Keywords

ClassCastException: Exception that is thrown to indicate that the code has attempted to cast an object of one data type to another.

Compilation: The translation of source code by a compiler into a lower level form.

JVM: Java Virtual Machine.

Methodology: The system of methods that is followed in a particular discipline.

OutOfMemoryError: Exception that is thrown when the JVM cannot allocate an object, since it is out of memory and no more memory can be made available by the garbage collector.

7.9 Self Assessment

1. State whether the following statements are true or false:
   (a) When an exception object is created and handed over to the run time system, then it is called catching an exception.

(b) Errors are usually ignored in program coding as hardly anything is done about an error.

(c) It is very much essential to avoid error-handling for creating a constraint application.

(d) Once the throw statement is encountered, the flow of execution stops immediately and none of the consequent statement executes further.

(e) The method, which is calling another method along with a throws clause, is required to be enclosed within the try/catch blocks.

(f) On closing and recovery of a file, it is essential to place the expression in the try block.

2. Fill in the blanks:

(a) A _____ is an exception that can be possibly avoided by the programmer.

(b) The _____ block holds some code, which might produce error.

(c) The _____ block includes the error handling code, which processes error declaring it as exception.

(d) The catch blocks that are employed immediately after the try block in which the exception was thrown, are denoted as __.

(e) The catch blocks employed at the higher level of the call hierarchy, projected to achieve universal error handling are denoted as _____.

3. Select a suitable option for every question.

(a) When the keyword **throws** is used along with a method, it is called as _____.

   (i) Ducking

   (ii) Throwable

   (iii) Threads

   (iv) Uncaught exception

(b) Which of the following methods describes the exception in program?

   (i) String toString( )

   (ii) String getMessage( )

   (iii) Throwable fillInStackTrace( )

   (iv) void print StackTrace( )

(c) Which exception is thrown, when an array that is not compatible with the data type of that array is declared?

   (i) Arithmetic Exception

   (ii) Class Cast Exception

   (iii) Array Index Out Of Bounds Exception

   (iv) Number Format Exception

(d) Which of the following is not an exception?

   (i) Array Store Exception

   (ii) Null Pointer Exception

   (iii) String getLocalizedMessage

   (iv) Negative ArraySizeException

(e) Which of the following is not a method?

    (i) String getLocalizedMessage

    (ii) Negative ArraySizeException

    (iii) void printStackTrace(PrintStream stream)

    (iv) void print StackTrace( )

7.10 Review Questions

1. "The **Exception** class does not define any of its own methods, but it does inherit the methods provided by **Throwable**". Comment.

2. "Inclusion of **try** block in your exception class is necessary". Elaborate.

3. "The checked exceptions cannot be thrown at a point where the error is unpredictable". Justify.

4. "The unchecked exception cannot be thrown at the time of compilation" Justify.

5. "When a **throw** statement is encountered, the further execution stops automatically". Discuss.

6. "There is difference between the functionalities of **throw** and **throws**". Comment.

7. Give the significance of the **String toString( )** method.

8. "Once the **throw** statement is encountered, the flow of execution stops immediately and none of the consequent statements are executed". Discuss.

9. "**Finally** blocks are executed once the **try** block exits". Justify?

10. "JVM (Java Virtual Machine) performs several tasks whenever an uncaught exception arises". Discuss.

11. "On closing and recovery of a file, the expression should be placed in the finally block". Justify?

Answers: Self Assessment

1. (a) False (b) True (c) False (d) True (e) True (f) False

2. (a) Runtime Exception (b) try (c) catch (d) Local exception handlers (e) Global exception handlers

3. (a) Ducking (b) String toString( ) (c) Array Index Out Of Bounds Exception (d) String getLocalizedMessage (e) Negative ArraySizeException

7.11 Further Readings

Books

Osborne McGraw-Hill, Java A Beginners Guide, 3rd Edition, Mar 2005.

Schildt. H. Java 2 The Complete Reference, 5th ed. New York: McGraw-Hill/Osborne.

Online link

http://www.cs.qub.ac.uk/~P.Hanna/JavaProgramming/Lecture8/Java%20-%20Lecture%208%20-%20Exceptions.pdf http://www.roseindia.net/java/java-exception/listofjavaexception.shtml

http://www.javamex.com/tutorials/exceptions/exceptions_uncaught_gui.shtml

http://www.java-samples.com/showtutorial.php?tutorialid=294

http://www.roseindia.net/help/java/t/throws-java-keyword.shtml

http://www.roseindia.net/java/java-exception/finally-exception.shtml

Unit 8: Multithreaded Programming I

CONTENTS

Objectives

After studying this unit, you will be able to:

- Analyze the life cycle of threads

- Describe the creation of threads

- Discuss the creation of multiple threads

Introduction

Java programming language has many important and useful concepts. One such concept is the **Thread**. A thread can be referred to as a single sequential flow of control within a program or the unit of execution within a process. A process is normally broken down into tasks and these tasks are further broken down into threads.

 Example: Consider the modern operating system, which allows multiple programs to run at once. While typing a document in a system, one can simultaneously listen to music and browse the net. This indicates that the operating system installed in the computer allows multitasking. Similarly, the execution of several processes in a program can also be done simultaneously. Hot Java web browser is an example of such an application, which allows the user to browse a Web page while downloading an image, or playing animations or audio files at the same time.

Java has a built-in support for threads. A larger portion of Java architecture is based on multi-threading. Threads in Java are used to allow an applet for accepting the input from user and simultaneously display the animations on the screen.

A thread has a beginning, a sequence of steps for execution, and an end. A thread is not considered as a program, but it runs within a program. Each and every program contains at least one thread called as primary thread. In Java, the **main( )** method is an example of a primary thread.

The memory for processes being executed is allocated by the microprocessor. Each and every process occupies its own memory or address space. And, all the threads in the process share the same address space. The thread is also called as **lightweight** process. Threads run in the same process space as the main program. Figure 8.1 shows the relationship between a thread and a process.

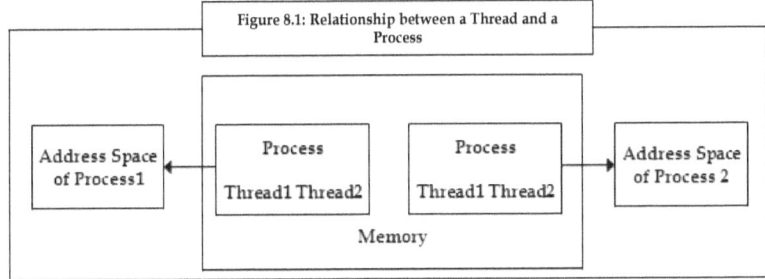

Figure 8.1: Relationship between a Thread and a Process

8.1 Threads

We know that in modern operating systems, we can execute multiple tasks under the interrupt driven operating system. This ability to handle more than one task at the same time is known as multitasking. In the system's terminology, it is referred to as multithreading.

A process made up of only one thread is called as a single-threaded process. Single threaded process performs only one task at a time whereas, a process having more than one thread, called as a multi-threaded process, performs different tasks and interacts with other processes at the same time. Figure 8.2 is the schematic representation of a single-threaded and a multi-threaded process.

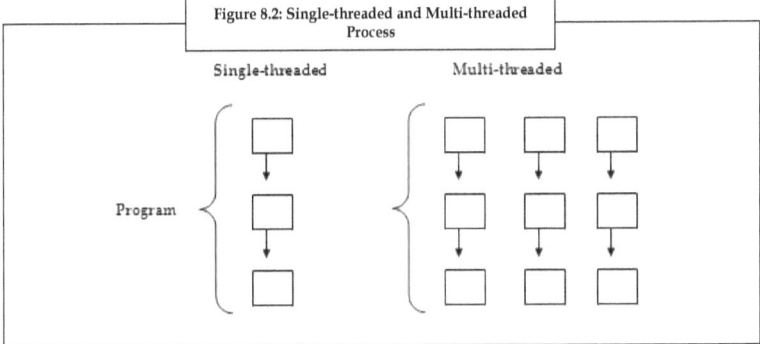

Figure 8.2: Single-threaded and Multi-threaded Process

In Java, **java.lang.Thread** class creates and controls each and every thread. A Java program has many threads, and these threads run either asynchronously or synchronously.

Multithreaded programming requires a different way of looking at the software. In this programming, many tasks are executed concurrently, that is, many tasks are performed at the same time. It is possible in multithreaded programming to start a new task even when the current task is not completed. Multithreading is a conceptual programming, where a program is divided into two or more subprograms, which can be implemented at the same time in parallel. Multithreading is also known as multiple-threads of execution. If the application performs many different tasks concurrently, then the threads may access shared data variables to work collaboratively.

8.2 Life Cycle of a Thread

Understanding the life cycle of a thread is very important, especially at the time of developing codes using threads. When a thread is alive, it indicates it is in one of its several states. However, invoking **start( )** method does not mean that the thread can access CPU and start executing immediately. There are several factors that determine this execution.

The life cycle of a thread is depicted in figure 8.3. This figure is not a complete diagram, but rather is an overview of the common phases of a thread's life.

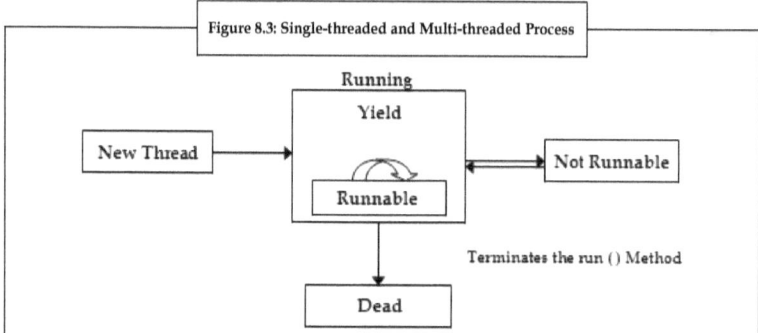

Figure 8.3: Single-threaded and Multi-threaded Process

As per the figure 8.3, the four possible states/phases of a thread's life cycle are:

1. New thread

2. Runnable

3. Not runnable

4. Dead

New Thread

When the **Thread** class is created at any instance, a thread enters into a new thread state.

Syntax of Creating a New Thread:

> Thread newThread1 = new thread(this);

As per the above syntax, no resources are allocated for the new thread, and **this** keyword denotes that the **run( )** method of the current object needs to be invoked.

Hence, it is an empty object. However, the **start( )** method needs to be invoked, to start the thread as given below:

> newThread.start( );

Runnable

In the **Runnable** phase of a thread's life cycle, the **start( )** method, which is responsible for starting the thread, allocates the system resources necessary for thread, schedules and plans the thread to run, and calls **run( )** method for the thread.

As soon as the **start( )** method of a thread in invoked; it causes the thread to move into the **Runnable** state. All the activities of the thread are carried out in the body of the thread, that is, the **run( )** method. As a single processor cannot perform the execution of more than one thread at a time; it maintains a thread queue. As soon as the threads are started, a thread queue is created for processor time, and the thread waits for its turn for execution. Therefore, for a given time, a thread waits for the processor's attention. Hence, the state of thread is defined as **runnable**.

Not Runnable

A thread is considered as not **runnable**, if its state is:

1. Sleeping

2. Waiting

3. Being blocked by another thread

The **sleep( )** method is used to put the thread into sleeping mode. A sleeping thread can enter into its **runnable** state, only after some specified time. Until and unless the specified time has elapsed, the thread will not be executed.

Syntax to Put a Thread into Sleeping Mode:

sleep(long s);

In the above syntax, **s** is the number of milliseconds for making a thread inactive.

Dead

The thread enters the dead state when the **run( )** method is complete, when an interrupt does not kill the thread, or when a destroy() method kills the thread but does not release the object locks of the thread.

 Example: If the loop in **run( )** method has fifty iterations, then the life of the thread will be fifty iterations of the loop.

A thread can be killed by assigning null to the thread object. The **isAlive( )** method in the Thread class determines whether the thread is started or stopped.

8.3   Creating a Thread

A thread that is created, must be bound to the **run( )** method of an object. At the start of a thread, it invokes the object's **run( )** method.

In Java, there are two ways (as given in the figure 8.4) to create threads, that is, by:

1. Implementing the **Runnable** Interface (java.lang.**Runnable**).

2. Extending the Thread Class (java.lang.Thread).

Figure 8.4 shows the methods used for creation of threads.

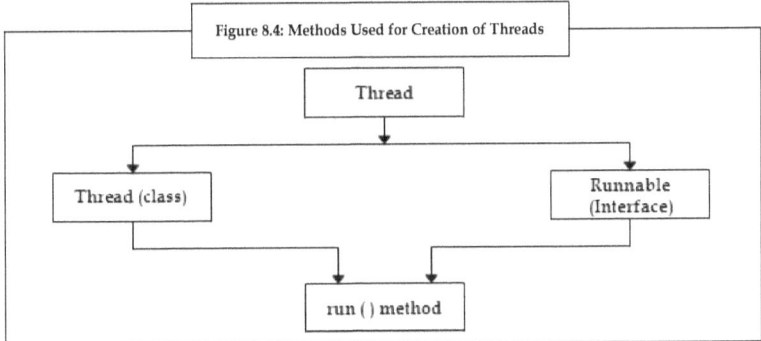

Figure 8.4: Methods Used for Creation of Threads

NAGESH JAITAK

Implementing the Runnable Interface

The easiest way to create a thread in Java, is by implementing the **runnable** interface, and then instantiating an object of that class. Using **run( )** method in the class, is the only method that is used for implementing the **runnable** interface, as this contains the logic of the thread.

Threads based on the **runnable** interface can be created by following the below steps:

1. The **runnable** interface is implemented by a class, providing the **run( )** method, which will be executed by the thread. An object of this class type is a **runnable object**.

2. The object of the Thread class is developed by passing a **runnable object** as an argument to the Thread constructor. Therefore, the Thread object now having a **runnable object** implements the **run( )** method.

3. The **start( )** method is now invoked on the Thread object, which is created in the previous step. And, the **start( )** method returns immediately after a thread has been produced.

4. The thread ends at the same time, when the **run( )** method ends.

 *Example:* Program to demonstrate the creation of new thread, by implementing the Runnable interface, and starting it to run.

```
//Create a second thread.
class MyNewThread2 implements Runnable
{
  Thread a;
  MyNewThread2( )
  {
   //Create a new, second thread a=new
   Thread(this, "A Thread");
   System.out.println("Demo thread:"+a);
   a.start( ); //Start the thread
  }
  //This is the entry point for the second thread.
  public void run( )
  {
   try
   {
    for(int j=5; j>0; j--)
    {
     System.out.println("Demo thread:"+j);
     Thread.sleep(1000);
    }
   } catch(InterruptedException e)
   {
```

```
System.out.println("Demo interrupted.");
}
System.out.println("Exiting Demo thread.");
}
}
class ThreadA
{
  public static void main(String args[ ])
  {
    new MyNewThread2( ); //create a new thread
    try
    {
      for(int j=5; j>0; j--)
      {
        System.out.println("The Main Thread:"+j);
        Thread.sleep(1500);
      }
    } catch(InterruptedException e)
    {
    System.out.println("The Main Thread interrupted.");
    }
    System.out.println("The Main Thread exiting.");
  }
}
```

Output:

Demo thread: Thread[A Thread, 5, main]

The Main Thread: 5

Demo Thread: 5

Demo Thread: 4

The Main Thread: 4

Demo Thread: 3

Demo Thread: 2

The Main Thread: 3

Demo Thread: 1

Exiting Demo Thread.

The Main Thread: 2

The Main Thread: 1

The Main Thread exiting

In this example,

1. First, a class **MyNewThread2** is created with **Runnable** interface using the **implement** keyword.

2. In this class, a thread **a** is declared.

3. Then, a constructor **MyNewThread2( )** of this class is called. In this constructor, a new thread is created by using the **new** keyword, and with the parameters **this** and **A Thread.** This new thread is linked to **a**. Then, the demo thread **a** is printed on the screen. Also, the **start( )** method of **a** is called.

4. The **run( )** method of the class is then called to move the thread into running state. In this method, try-catch mechanism is used.

   (a) In the **try** block,

      (i) A **for** loop is used, wherein the value of integer **j** is initialized to 5 and the condition j>0 is checked.

      (ii) If the condition is true, then the statement **System.out.println** (**"Demo thread:"+j**) is used to print the Demo thread **j**. The thread is made to sleep using the **Thread.sleep(1000)** statement.

      (iii) Then the value of j is decremented by 1.

      (iv) Steps (i) to (iii) are repeated till the condition in step (i) is true.

   (b) In the catch, **e** is declared as the InterruptedException. In the catch block, **System.out.println("Demo thread:"+j)** is used to print Demo interrupted.

   (c) After the catch block, **System.out.println("Exiting Demo thread.")** statement is used to print Exiting Demo thread.

5. A new class ThreadA is defined, wherein the main( ) method of the class is called. In this main( ) method,

   (a) The constructor MyNewThread2( ) is called using the new operator.

   (b) In the try block,

      (i) A for loop is used, wherein the value of integer j is initialized to 5 and the condition j>0 is checked.

      (ii) If the condition is true, then the statement System.out.println("The Main Thread:"+j) is used to print the Demo thread j. The thread is made to sleep using the Thread.sleep(1500) statement.

      (iii) Then the value of j is decremented by 1.

      (iv) Steps (i) to (iii) are repeated till the condition j > 0is true.

   (c) In the catch block, e is declared as the InterruptedException. In the catch block, System.out.println("The Main Thread interrupted.") is used to print Demo thread.

   (d) After the catch block, System.out.println("The Main Thread exiting.") statement is used to print Exiting Demo thread.

Extending the Thread Class

The second way   used to create a thread is by creating a new class that extends Thread, and then creating an instance of that class. The extending class overrides the **run( )** method, acting as an entry point for creating a new thread.

New threads based on extending the Thread class can be created by following the below given steps:

1. A class being extended from the Thread class overrides the **run( )** method from that Thread, for defining the code to be executed by the thread.

2. This subclass further calls a Thread constructor in its constructors, for initializing the thread. This is done by using the **super( )** call.

3. In order to make the thread eligible for running, the **start( )** method is invoked on the object of the class.

 Example: Program to illustrate the instantiation and running of threads by extending the Thread class.

```
class NewThreadY extends Thread
{
  NewThreadY( )
  {
  super("DemoA Thread");
  System.out.println("ChildC thread:"+this);
  start( );
  }
  public void run( )
  {
  try
  {
    for(int j=5; j>0; j--)
    {
      System.out.println("ChildC Thread:"+j);
      Thread.sleep(400);
    }
  }catch(InterruptedException e)
  {
    System.out.println("ChildC interrupted.");
  }
  System.out.println("Exiting ChildC thread.");
  }
}
class ExtendThread
```

```
{
public static void main(String args[ ])
{
new NewThreadY( );
try
{
  for(int j=5; j>0; j--)
  {
   System.out.println("The Main Thread:"+j);
   Thread.sleep(1500);
  }
} catch(InterruptedException e)
{
   System.out.println("The Main Thread interrupted.");
}
   System.out.println("The Main Thread exiting.");
}
}
```

Output:

ChildC Thread: Thread[DemoA Thread, 5, main]

The Main Thread: 5

ChildC Thread: 5

ChildC Thread: 4

The Main Thread: 4

ChildC Thread: 3

ChildC Thread: 2

The Main Thread: 3

ChildC Thread: 1

Exiting ChildC Thread.

The Main Thread: 2

The Main Thread: 1

The Main thread exiting.

In this example,

1.   First, a class **MyNewThread2** is created using the keyword **extends**.

  (a) Then the constructor **NewThreadY( )** is declared.

  (b) The call to **super( )** inside **NewThreadY** invokes the Thread constructor in the following form:

public Thread(String threadName)

Here, **threadName** is used to specify the name of the thread.

(c) The statement System.out.println("ChildC thread:"+this) prints ChildC thread pointed by this keyword.

(d) Then the start( ) method is called to start the thread.

2. The run( ) method of the class is then called to move the thread into running state. In this method, the try-catch mechanism is used.

(a) In the **try** block,

(i) A for loop is used, wherein the value of integer j is initialized to 5 and the condition j>0 is checked.

(ii) If the condition is true, then the statement System.out.println("ChildC Thread:"+j) is used to print the Demo thread j. The thread is made to sleep using the Thread.sleep(400) statement.

(iii) Then the value of j is decremented by 1.

(iv) Steps (i) to (iii) are repeated till the condition j>0 is true.

(b) In the catch block, e is declared as the InterruptedException. In the catch block, System.out.println("ChildC interrupted.") is used to print Demo interrupted.

(c) After the catch block, System.out.println("Exiting ChildC thread.") statement is used to print Exiting Demo thread.

2. A new class ExtendThread is defined, wherein the main( ) method of the class is called. In this main( ) method,

(a) A constructor **MyNewThreadY( )** is called using the **new** operator.

(b) In the try block,

(i) A for loop is used, wherein the value of integer j is initialized to 5 and the condition j > 0 is checked.

(ii) If the condition is true, then the statement System.out.println("The Main Thread:"+j) is used to print the Demo thread j. The thread is made to sleep using the Thread.sleep(1500) statement.

(iii) Then, the value of j is decremented by 1.

(iv) Steps (i) to (iii) are repeated till the condition is true.

(c) In the catch block, e is declared as InterruptedException. In the catch block, System.out.println("The Main Thread interrupted.") is used to print Demo thread.

(d) After the catch block, System.out.println("The Main Thread exiting.") statement is used to print Exiting Demo thread.

8.4   Creating Multiple Threads

So far, we have been using only two threads, namely, one main thread and one child thread. Let us see an example that uses multiple threads.

Example: Program to illustrate the concept of multiple threads.

```
//create multiple threads
class NewThread implements Runnable
{
String name; //name of thread
Thread t;
NewThread (String threadname)
{
name= threadname;
t=new Thread (this, name)
{
System.out.println("New thread:"+t);
t.start( ); //start the new thread
}
//this is the entry point for thread
public void main( )
{
try{
for(int i=5; i>0; i--)
{
System.out.println(name + ": "+i);
Thread.sleep(1000);
}
}
catch (InterruptedException e)
{
System.out.println(name + "Interrrupted");
}
System.out.println(name + "exiting");
}
}
class MultiThreadDemo
{
public static void main(String args [ ])
```

```
{
new NewThread("One");
new NewThread("Two");
new NewThread("Three");
try
{
//wait for other threads to end
Thread.sleep(10000);
}
catch (InterruptedException e)
{
System.out.println("Main thread Interrrupted");
}
System.out.println("Main thread exiting");
}
}
```

Output:

```
New thread: Thread[One, 5, main]
New thread: Thread[Two, 5, main]
New thread: Thread[Three, 5, main]
One:5
Two:5
Three:5
One:4
Two:4
Three:4
One:3
Two:3
Three:3
One:2
Two:2
Three:2
One:1
Two:1
Three:1
One exiting
```

Two exiting

Three exiting

Main Thread exiting

In this example,

1. First, a class **MyNewThread2** is created with **Runnable** interface using the **implements** keyword. In this class,

    (a) A string **name**, and thread **t** are declared.

    (b) Then, a constructor **NewThread( )** of this class is called for the **threadname**. This constructor is used for linking the **threadname** to **name**.

    (c) A constructor **new Thread( )** is called with the parameters **this** and **name**. This is linked to thread **t**. Thereafter, the statement **System.out.println("New thread:"+t)** prints the **New thread t**. Also, the start( ) method is called to start the thread **t**.

2. In the **main( )** method, **try-catch** mechanism is used.

    (a) In **try** block,

    (i) A for loop is used, wherein the value of integer i is initialized to 5 and the condition i>0 is checked.

    (ii) If the condition is true, then the statement System.out.println(name + ": "+i) is used to print the Demo thread i. The thread is made to sleep using the Thread.sleep(1000) statement.

    (iii) Then, the value of i is decremented by 1.

    (iv) Steps (i) to (iii) are repeated till the condition i>0 is true.

    (b) In the catch block, e is declared as the **InterruptedException**. In the **catch** block, **System.out.println(name + "Interrrupted")** is used to print name, that is, Interrupted.

    (c) After the catch block, **System.out.println(name + "exiting")** statement is used to print the name of the thread that is **exiting.**

3. A class **MultiThreadDemo** is defined.

    (a) Three child threads **One**, **Two** and **Three** are created.

    (b) After this **try-catch** block mechanism is used.

    (i) In try block, the thread is made to sleep using *Thread.sleep(10000)*.

    (ii) In the catch block, e is declared as the InterruptedException. In the catch block, *System.out.println("Main thread Interrrupted")* is used to print Main thread Interrrupted.

    (iii) After the catch block, *System.out.println("Main thread exiting")* statement is used to print *Main thread exiting*.

8.5 Summary

- A thread can be referred to as a sequential flow of control within a program or the unit of execution within a process.

- The Thread class is defined in the pre-defined package **java.lang**, which needs to be imported in the program code, so that our classes are aware of their definition.

- Multithreading is the ability of an operating system to execute different parts of a program simultaneously.

- There are four possible states in the life cycle of a thread, namely, the **new thread** state, the **runnable** state, the **not runnable** state and the **dead** state.

- There are two ways to create a thread. First, by implementing the **Runnable** interface and second, by extending the Thread class.

8.6 Keywords

CPU: Central Processing Unit.

Java Messaging System: Provides a consistent API set that gives developers access to the common features of many messaging systems products.

Processor: Circuit that executes computer programs.

run( ): Method used to implement the code that needs to be executed by our thread.

start( ): Method that causes the thread to move into the **Runnable** state.

8.7 Self Assessment

1. State whether the following statements are true or false:

    (a) A thread moves to the running state once it has been started.

    (b) Threads can be created by either extending the Thread class or by implementing the Runnable interface.

    (c) The start( ) method can be called only once in the life cycle of a thread.

2. Fill in the blanks:

    (a) You can create a new thread by extending the class _____.

    (b) The extending class overrides the _____ method.

    (c) The easiest way to create a thread in Java, is by implementing the _____ interface, and then instantiating an object of that class

3. Select a suitable option in the following questions.

    (a) What does sleep in Thread class do?

        (i) Causes the thread, which sleep is invoked on, to sleep (temporarily cease execution) for the specified number of milliseconds

        (ii) Causes the currently executing thread to sleep (temporarily cease execution) for the specified number of milliseconds

        (iii) Causes the main( ) thread to sleep for the specified number of milliseconds

        (iv) Causes the currently executing thread to wait (temporarily cease execution) for

(b) Which state does the thread enter, when a thread class is created at any instance?

    (i) New thread

    (ii) Runnable

    (iii) Not Runnable

    (iv) Dead

8.8 Review Questions

1. "A thread that is created, must be bound to the run( ) method of an object". Comment.

2. "When a thread is alive, it indicates it is in one of its several states". Justify.

3. "The second way that is used to create a thread is, by creating a new class that extends Thread, and then creating an instance of that class". Elaborate.

4. "Understanding the life cycle of a thread is very important, especially at the time of developing codes using threads". Elaborate.

Answers: Self Assessment

1.   (a) False

    (b) True

    (c) True

2.   (a) Thread

    (b) run( )

    (c) runnable

3.   (a) Causes the currently executing thread to sleep (temporarily cease execution) for the specified number of milliseconds.

    (b) New Thread

8.9 Further Readings

Books

E Balagurusamy, Programming with Java_A Primer 3e, New Delhi

Herbert Schildt, The Complete Reference, 7th edition, Tata McGraw Hill

Online link

http://publib.boulder.ibm.com/infocenter/cicsts/v3r1/index.jsp?topic=/com.ibm.ci
cs.ts31.doc/dfhpj/topics/dfhpj_thread.htm

http://java.sun.com/products/jfc/tsc/articles/threads/threads1.html

Unit 9: Multithreaded Programming II

Objectives

After studying this unit, you will be able to:

- Explain the Java thread model
- Describe stopping and blocking a thread
- Describe inter-thread communication
- Explain suspending, resuming and stopping a thread

Introduction

A multithreaded application delivers its potent power by concurrently running several threads in a single program. Logically, multithreading refers to the execution of multiple lines of a single program at the same time. However, it is not the same as starting a program twice and saying that there are multiple lines of a program being executed at the same time. Here, the operating system is treating the programs as two separate and distinct processes.

Multithreaded programming requires a different method of dealing with software. That is, instead of executing a series of steps sequentially, tasks are executed concurrently, rather than waiting for one task to finish before starting another. In the previous unit, we learnt the different stages involved in the life cycle of a thread. We also learnt to create a single thread and to create multiple threads. In this unit we shall learn more about multithreaded programming.

9.1 The Java Thread Model

In Java runtime machine, all the class libraries are designed with multithreading in mind. The Java runtime machine depends on threads for many things like sharing the memory space and code. Java uses threads to enable the entire Java runtime environment to be asynchronous. This reduces inefficiencies like consumption of more CPU cycles or wastage of CPU time by preventing the wastage of CPU cycles.

The value of multithreaded environment is best understood in contrast to its counterpart (Single threaded environment). The single thread systems use the event loop approach with polling. In Java thread model, a single thread polls a single event queue to decide what to do next. If the polling mechanism returns with a signal that a network file is ready to be read, then the event loop dispatches control to the appropriate event handler. Any other action cannot be performed, until this event handler returns. This wastes CPU time. This can also appear only in one part of the program dominating the system and preventing any other events from being processed. In a single threaded environment, if a thread is blocked, it means that it is waiting for some resources. During this polling mechanism, the entire program is stopped. Hence, this polling mechanism is eliminated in Java's multithreading. Here, one thread can pause without stopping other parts of your program.

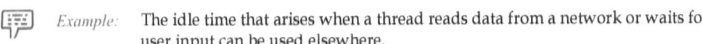 *Example:* The idle time that arises when a thread reads data from a network or waits for user input can be used elsewhere.

In animation programs, multithreading allows the loops to sleep for a second between each frame without causing the complete system to pause. If a thread is blocked, only the single thread, that is blocked, pauses. The remaining threads continue to run.

Thread exists in many states. A thread can exist in running state. A running thread can be resumed, suspended or blocked. At any point of time, the thread can also be terminated, which stops its execution immediately. A thread cannot be resumed, once it is terminated.

9.1.1 Thread Priorities

As discussed earlier, multithreading is achieved by switching between one thread and another to create concurrent execution of code. If the program execution has to happen with multiple CPUs, then the operating system will move between one thread and another based on the arbitrary algorithm. So, this makes it impossible to predict the order in which threads will be executed. The best way to predict the execution of thread is assigning the relative priority for threads. This enables the operating system to determine which thread is important.

Thread priorities are integer values that indicate the relative priority of one thread to another. But, when the thread priority is considered as an absolute value, it is meaningless. The thread with higher priority does not mean it runs faster than the thread with lower priority. Instead, the thread priority helps to decide when to switch from one running thread to the next. This is referred to as a context switch. The rules of context switch are:

1. A thread can voluntarily surrender control. This can be achieved by yielding, sleeping, or blocking on pending I/O. In this situation, all the other threads are examined, and the thread with higher priority is given the CPU time.

2. A thread can be preempted by a higher priority thread. In this situation, a lower priority thread does not yield the processor but is simply preempted. Always the thread with higher priority runs first. This process is called preemptive multitasking.

Assigning a Thread Priority

In Java, numerical ranking symbolizes the thread priority, where 10 is the highest priority and 1 is the lowest priority. Some thread priorities are defined as static member variables of the **java.lang**. The method **Thread.SetPriorityMehtod(int)** is used to set the thread priority.

Example: Thread t=new Thread (**runnable**);

t.setPriority (Thread.MIN_PRIORITY);

t.start ( );

This example is used to set the minimum priority level.

Table 9.1 depicts the thread priorities.

Table 9.1: Thread Priorities	
10	Thread.MAX_PRIORITY
9	
8	
7	
6	
5	Thread.NORM_PRIORITY
4	
3	
2	
1	Thread.MIN_PRIORITY

Obtaining the Current Thread Priority

If the thread wishes to determine its current thread priority, then it can do so, by invoking the **Thread.getPriority( )** method. An **int** value is returned, which indicates the priority of the thread.

 Example: Thread t = Thread.currentThread( );
System.out.println ("Priority:"+t.getPriority( ));

In this example, **Thread.currentThread( )** method is used to get the priority of the currently running thread.

Limiting Thread Priority

Sometimes it becomes necessary to limit the maximum priority of a thread. This can be made possible by installing a custom security manager, which throws a **SecurityException**. But, this process of installing a custom security manager involves a significant amount of effort. An easier approach is to create a **ThreadGroup** and assign a maximum priority level to this group. Thread group is specified while creating a thread. **ThreadGroup.setMaxPriority(int)** method is used to assign maximum thread priority for a group.

 Example: ThreadGroup group = new ThreadGroup ("mygroup");
group.setMaximumPriority(8);

In this example, the thread priority is set to 8.

Example: Program to illustrate thread priorities.

```
/* class Clicker is declared with Runnable interface using implement keyword */
class clicker implements Runnable
{
int clicker = 0;        //clicker is assigned value of 0
Thread t;               //thread t is created
private volatile Boolean running = true;
public clicker (int p)  //Clicker method is called with integer p
{
t=new Thread(this);     //a new thread is created and linked to t
t.setPriority(p);       //Priority of the thread is set
}
public void run( )      //run( ) method is invoked
{
while(running)          //value of running is examined using while loop
{
click++;
}
}
public void stop( )     //stop( ) method is invoked
{
running = false;
}
public void start( )    //start( ) method is called
{
t.start( );             //thread t is started
}
}
class ThePrioDemo       //class ThePrioDemo is declared
{
public static void main(String args[ ])
{
Thread.currentThread( ).setPriority(Thread.MAX_PRIORITY);
//priority is set two values above the normal priority
clicker the = new clicker(Thread.NORM_PRIORITY+2);
//priority is set two values below the normal priority
```

```
clicker pri = new clicker(Thread.NORM_PRIORITY-2);
the.start( );          //thread is started
pri.start( );          //thread is started
try
{
Thread.sleep(10000);        //thread is made to sleep for ten seconds
}
catch(InterruptedException e)
{
System.out.println("Main thread interrupted");
}
the.stop( );                    //thread is stopped
pri.stop( );                    //thread is stopped
try
{
the.t.join( );                  //threads are joined
pri.t.join( );                  //threads are joined
}
catch(InterruptedException e)
{
System.out.println("InterruptedException caught");
}
System.out.println("Low priority thread:"+the.click);
System.out.println("High priority thread:"+pri.click);
}
}
```

Output:

Low priority thread: 4408112

High priority thread: 589626904

In this example, two threads at different priorities run on the same platform.

One thread is set to normal priority, as defined by **Thread.NORM_PRIORITY**, and the other thread is set to two levels below it. Then the **run( )** method is called, which starts the thread and allows it to run for ten seconds. Each thread executes a loop, counting the number of iterations. Here, running is preceded by the keyword **volatile**. This keyword ensures that the value of running is examined each time the following loop iterates:

while(running)

```
{
click++;

}
```

This keyword optimizes the loop such that a local copy of running is created.

The exact output produced by the above program depends upon the speed of your CPU. The output also depends on the number of other tasks running in the system.

9.1.2 Synchronization

Threads run concurrently and are independent of each other. This indicates that the threads run in their own space, without being concerned about the status and activities of the other threads that are running concurrently. During this process, threads do not require any method or outside resources, and as a result, threads do not communicate with each other. These types of threads are generally called as **asynchronous threads**. Sharing the same resource (variable/method) by two or more threads is the major problem suffered by asynchronous threads, as only one thread can access a resource at a time.

Let us assume that two threads Thread1 and Thread2 are the producer and the consumer processes respectively. These two threads share the same data. A situation can arise in which either the producer produces the data faster than it is consumed or the consumer retrieves the data faster than it is produced. This problem is schematically represented in figure 9.1.

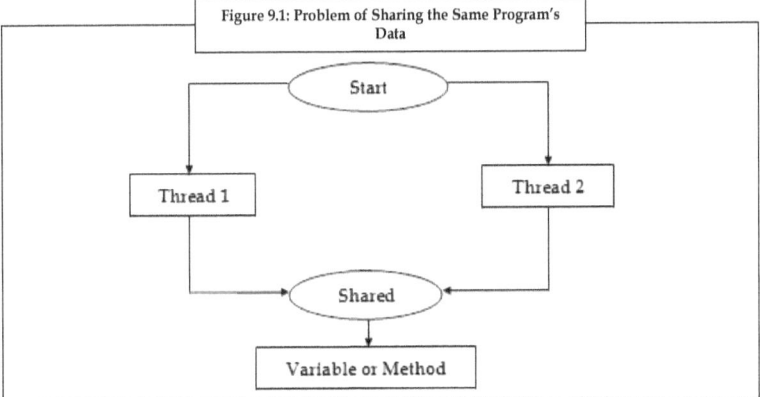

To avoid and to solve the above problem, Java uses a monitor, which is commonly known as a **semaphore**. This prevents the data from being corrupted by multiple threads. We can implement a monitor by using the keyword **synchronized** to synchronize threads so that they can intercommunicate with each other. This mechanism allows two or more threads to share the resources in a sequential manner. Java's **synchronized** keyword ensures that only one thread at a time is in a critical region. This region is a locked area, where only one thread at a time is run (or locked). Therefore, once the thread is in its critical region, no other thread can enter into that critical region, and the thread has to wait until and unless the current thread leaves its critical region.

Syntax for Synchronized Statement

synchronized (object)

{

//statements to be synchronized

}

Locking an Object

The term **lock** refers to the access granted or approved to a particular thread to access the shared resources. To ensure that only one thread can access a method, Java locks an object, including its methods. When an object is in the monitor, it makes sure that no other thread can access the same object. Java object has a built-in lock system, which is enabled only when the object has synchronized method code. An object acts as a guard in ensuring the synchronized access to the resource, by associating the shared resource with a Java object and its lock. Therefore, only one thread can access the shared resource guarded by the object lock at a time. There is only one lock per object, and if one thread acquires the lock, no other thread can acquire it until the first thread releases the lock.

Syntax for Synchronizing an Object

synchronized (<object>)

{

//statements to be synchronized

}

By using the above syntax, the methods of the object can only be invoked by one thread at a time.

There are two ways, in which, the synchronization of the execution of code can be done. These two ways are:

1. Synchronized Methods

2. Synchronized Blocks (Statements)

Synchronized Methods

If any method is specified with the keyword **synchronized**, it will be executed by only one thread at a time. For any thread to execute the **synchronized** method, first, it has to obtain the objects lock. However, if the lock is held by another thread, then the calling thread has to wait. These methods are useful in situations where different methods need to be executed concurrently, so that these methods can intercommunicate and manipulate the state of an object.

 Example: Program to illustrate synchronized method.

```
class share extends Thread      //class share is declared using keyword extends
{
 static String msg[ ]={"Following", "is", "a", "synchronized", "variable"}; //msg is
 declared as static
 share(String threadname)
 //share is called with the parameter threadname
 {
  super(threadname);              //super class is declared
 }
```

```
public void run( )                    //run( ) method is invoked
{
  display(getName( ));                //displays the name of the thread
}
public synchronized void display(String threadN)
{
for(int k=0; k<=4; k++)
  System.out.println(threadN+msg[k]);
  try
  {
  this.sleep(1500);                   //thread is made to sleep
  } catch(Exception e) { }
  }
  }
  }
public class SynchroThread1
{
 public static void main(String[ ] args)
{
share T1=new share("Thread 1:");          //a new thread is created
T1.start( ); //thread is started
share T2=new share("Thread 2:");          //a new thread is created
T2.start( );//thread is started
}
} Output:
SynchroThread1
Thread 1: Following
Thread 1: is
Thread 1: a
Thread 1: synchronized
Thread 1: variable
Thread 2: Following
Thread 2: is
Thread 2: a
Thread 2: synchronized
Thread 2: variable
```

In this program, the **display( )** method is synchronized, and is shared by both the thread's objects at the same time during program execution. Therefore, only one thread is able to access the method and to process it until that method's statements are executed.

The following program illustrates the synchronization of the print method in TwoString class.

 Example: Program to illustrate synchronization of the print method.

```
class TwoStrings  //class TwoString is declared
{
//the print method is declared with two parameters that are of type String
synchronized static void print(String strg1, String strg2)
{
System.out.print(strg1);
try
{
 Thread.sleep(400);   //thread is made to sleep
} catch (InterruptedException e) {}
 System.out.println(strg2);
}
}
class PrintStringsThread implements Runnable
{
Thread thread;
String strg1, strg2;
PrintStringsThread(String strg1, String strg2)
{
 this.strg1 = strg1; // this keyword is used to refer to an instance of the class
from its method
 this.strg2 = strg2;   // this.strg1 and this.strg2 refers to the instance variable
strg1 and strg2, were strg1 and strg2 refers to the arguments passed in the
method PrintStringsThread
 thread = new Thread(this);
 thread.start( );
}
public void run( )
{
TwoStrings.print(strg1, strg2);
```

```
        }
        }
        class TestThread
        {
        public static void main(String args[ ])
        {
        new PrintStringsThread("Hello ", "Ma'am."); new
        PrintStringsThread("How do ", "you do?"); new
        PrintStringsThread("Nice ", "meeting you.");
        }
        }
```

Output:

Hello Ma'am.

How do you do?

Nice meeting you.

In this program, the "TwoStrings.print(strg1, strg2)" is synchronized with "PrintStringsThread(String strg1, String strg2)". Different values are passed as the string parameters and the program calls the Twostrings.print() method to print the output in the desired format.

Task

Write a program to illustrate the synchronization of the print method that displays the following output:

My name is John

I am a student of Nagesh Jaitak .

Synchronized Blocks (Statements)

An alternative way of handling synchronization is by using Synchronized Blocks (Statements).

Syntax for Synchronized Block

```
        synchronized (object reference expression)
        {
        //statements to be synchronized
        }
```

These synchronized statements must specify the object providing the native lock.

Example: Program to illustrate a synchronized object.

```
class Share extends Thread
//class Share is declared using keyword extends
{
static String msg[ ]={"Following", "is", "a", "synchronized", "variable"}; //msg is
declared static
Share(String threadname)      //share is called with parameter threadname
{
 super(threadname);           //super class is declared
}
 public void run( )           //run( ) method is invoked
 {
  display(getName( ));        //displays the name of the thread
 }
 public void display(String threadN)
 {
 synchronized(this)          //synchronized block is declared
 {
  for(int j=0; j<=4; j++)
  System.out.println(threadN+msg[j]);
  try
  {
   this.sleep(500);          //thread is made to sleep for 500 milliseconds
  } catch(Exception e) { }
 }
 }
}
public class SynchroStatement    //class SynchroStatement is declared
{
public static void main(String[ ]args)
{
 Share T1=new Share("Thread 1:");  //new thread is created
 T1.start( );                      //thread is started
 Share T2=new Share("Thread 2:");  //new thread is created
 T2.start( );                      //thread is started
}
```

}

Output:

SynchroStatement

Thread 1: Following

Thread 1: is

Thread 1: a

Thread 1: synchronized

Thread 1: variable

Thread 2: Following

Thread 2: is

Thread 2: a

Thread 2: synchronized

Thread 2: variable

In this program, the **synchronized (this)** is synchronized, and is shared by both the thread's objects at the same time during program execution. And, only one thread has the criteria to access the method and to process it until that method's statements are executed.

9.1.3 Messaging

After the division of a program into different threads, we need to define how they will communicate with each other. Java provides a clean and low cost platform for two or more threads to communicate with each other. This can be performed through the medium of calls to predefined methods that all objects have. Java's messaging system first allows a thread to enter a synchronized method on an object. Then, it waits there until some other thread explicitly notifies it to come out.

9.1.4 Thread Class and Runnable Interface

The Thread class in Java (java.lang.Thread class) is used for constructing and accessing individual threads in a multi-threaded application. Thread class supports a number of methods that includes information about all the activities of a thread; it sets and checks the properties of the thread, and it causes a thread to wait, be interrupted or be destroyed. By extending the Thread class, the applications and classes can be made to run in separate threads.

Syntax of Thread Class

    public class <class_name> extends Thread

 Example: public class DemoThread extends Thread
    {   ....
    }

In this example, a thread class **DemoThread** is created from the class **Thread**, and is declared **public**.

Many constructors are used in the Thread class; some of the important constructors are given in table 9.2.

Table 9.2: Thread Constructors

Thread Constructor	Description
Thread( )	Used to create a new Thread object.
Thread(String name)	Used to create a new Thread object with a specified name.
Thread(**Runnable** target)	Used to create a new Thread object on the basis of **Runnable** object. Target refers to the object whose run method is called.
Thread(**Runnable** target, String name)	Used to create a new Thread object with a specified name on the basis of a **Runnable** object.

Methods are as important as constructors in a Thread class. Some important methods provided by the Thread class are given in table 9.3.

Table 9.3: Methods Used in java.lang.Thread

Method	Return Type	Description
currentThread( )	Thread	Returns an object with reference to the thread in which it is invoked.
getName( )	String	Retrieves the name of the thread object or instance.
start( )	void	Calls run( ) method to initiate the thread. Starts the thread by calling its run( ) method.
run( )	void	Is the entry point for thread execution.
sleep( )	void	Suspends a thread for a specified time period (in milliseconds).
isAlive( )	Boolean	Determines whether the thread is running or not.
activeCount( )	int	Returns the active threads to its thread group and all its subgroups.
interrupt( )	void	Interrupts the thread on which it is being invoked.
yield( )	void	Allows the current thread's execution to pause temporarily and allows the other threads to execute.

Runnable Interface

We know that the **runnable interface** defines the method **run( )** that must be implemented. This method is implemented by classes to show that they are capable of being run as a separate thread of execution.

Syntax for run( ) Method:

        public void run( )

In this syntax, **run( )** method is called, and is declared **public** and **void**.

There are many advantages of using the **runnable interface**. These advantages are:

1. Object can be inherited from a different class.

2. The same runnable object can be passed to more than one thread (many threads can use the same code and also act on the same data).

3. Some applications can minimize the overhead, because some new thread instances require valuable memory and CPU time.

 *Example:* Program to create new threads with the **Thread** class and runnable objects by using classes that implement the **Runnable** interface.

```
class SampleRunnable implements Runnable
{
  public static void main(String[ ] args)
  {
    SampleRunnable SRun = new SampleRunnable( );
    Thread thr = new Thread(SRun);
    thr.start( );
    System.out.println("Hello world >> from the main program");
  }

  public void run( )
  {
    System.out.println("Hello world >> from a thread");

    try
    {
      Thread.sleep(1000*50*50);
    }

    catch (InterruptedException e)
    {
      System.out.println("Thread Interrupted");
    }
```

}

}

Output:

Hello world >> from the main program

Hello world >> from a thread

In the above program:

1. The **Runnable** interface is implemented in the class **SampleRunnable**.
2. A runnable object is instantiated.
3. A thread object of the **Thread** class is instantiated with the specified runnable object.
4. The **start( )** method is called on the new object.
5. The **run( )** method is implemented by defining it in the Runnable interface.

Note: The Thread object **thr** is created using the special **Thread** constructor, which takes a Runnable object as input. If a Thread object is created in this way, the **run( )** method of the **Runnable** object is executed as a new thread.

The **sleep( )** method has to be called explicitly by prefixing the class name: Thread, because the class is not extending the **Thread** class.

Write a program to illustrate a multi-threaded application that uses the runnable interface rather than a subclass of the Thread class.

Analyze the following code segment and find out what will happen when this code is executed?

Thread t =new Thread( );

t.start( );

t = null( );

9.2 Stopping and Blocking a Thread

At a certain point of time, it is necessary to terminate a thread before its task has been completed.

 Example: If the network client wants to send a message to a mail server in the second thread, then the thread in the execution state should be stopped immediately.

Thread can be stopped by another thread by invoking the Thread.stop( ) method. But this requires a controlling thread to maintain a reference to the thread that it wants to shut down.

Example: Program to illustrate the concept of stopping a thread.

```
//class StopMe is declared using keyword extends
public class StopMe extends Thread
{
public void run( )      //Run method is executed when thread first started
{
int count=1;           //count is assigned to 1
System.out.println("I can count");
for(;;)                //for loop is iterated
{
//Print count and increment it
System.out.print(count++ +" ");
try
{
Thread.sleep(500);    //Sleep for half a second
}
catch(InterruptedException ie)
{}
}
//Main method to create and start threads
public static void main(String args[ ]) throws java.io.IOException
{
//Create and start counting thread
Thread counter = new StopMe( );
counter.start( );
//Prompt user and wait for input
System.out.println("Press any enter to stop the thread counting");
System.in.read( ):
//Interrupt the thread
counter.stop;
}
```

In this example,

The thread will display an incrementing **count**, which will go on indefinitely without terminating. The **run( )** method is executed initially, when the thread is first started. Then the **count** gets incremented and also gets printed. Then the thread is made to **sleep** for half a second. After this, the **main( )** method creates and starts the thread. This starts counting the thread. Finally, the **Thread.stop( )** is used to stop the thread.

Thread can be blocked or suspended from getting into the **runnable** and next running states, by using either of the following thread methods:

sleep( )	//blocked for a specified time
suspend( )	//blocked until further orders
wait( )	//blocked until certain conditions occur

These methods affect the threads to go into the blocked state. In case of sleep( ) method, the thread method will return to the **runnable** state when the time is elapsed; in case of suspend( ) method, resume( ) method is invoked; in case of wait( ) method, notify( ) method is called.

9.3 Inter-thread Communication

There are three ways for the threads to communicate with each other.

In the first way, all the threads share the same memory space. If the threads share the same object, then these threads share access to that object's data member and thus communicate with each other.

In the second way, threads communicate by using thread control methods. The second way includes the following three methods by which threads communicate with each other:

1. *suspend( ):* By using this method, a thread can suspend itself and wait till another thread resumes it.

2. *resume( ):* By using this method, a thread can wake up another waiting thread through its resume( ) method and then run concurrently.

3. *join( ):* By using this method, the caller thread can wait for the completion of the called thread.

In the third way, threads communicate by using the following three methods:

1. *wait( ):* This method tells the calling thread to examine and make the calling thread wait until another calls the same thread's notify( ) or notifyAll( ) or a timeout occurs.

2. *notify( ):* This method wakes only the first waiting thread on the same object.

3. *notifyAll( ):* This method wakes up all the threads that have been called by wait( ) on the same object.

  Example: Program to illustrate the implementation of **wait( ), notify( ) and notifyAll( )** methods.

```
class Sort                //class Sort is created
{
int num=0;//interger value num is assigned 0
Boolean value = false;    //The variable value is assigned false
synchronized int get( )   //synchronized block is declared
{
 if (value==false)        //checking if value is false
 try {
  wait( );                //thread is made to wait
  }
```

```
        catch (InterruptedException e) {
        System.out.println("InterruptedException caught");
          }
        System.out.println("consume: " + num);
        value=false;
        notify( );              //notify method is invoked
        return num;             //returns num value
        }
        synchronized void put(int num)
        {
          if (value==true)   // checking if value is true
          try {
            wait( );             //thread is made to wait
            }
          catch (InterruptedException e) {
          System.out.println("InterruptedException caught");
            }
            this.num=num;      //num value points to the current class object
            System.out.println("Produce: " + num);
            value=false;
            notify( );          //notify method is invoked
            }
            }


        //class Construct is declared using keyword extends
        class Construct extends Thread
        {
          Sort s;              //sort is declared with object s (Synchronized method)
          Construct(Sort s)
        {
            this.s=s;          //s points to the current class objects
            this.start( );
          }
          public void run( )    //run( ) method is invoked
        {
            int i=0;            //integer i is assigned to value 0
            s.put(++i);         //increments the value num
```

```
     }
}
class User extends Thread   //class User is declared using extends keyword
{
  Sort s;                //sort is declared with object
  User(Sort s)           //constructor is called with Sort s (Synchronized method)
{
    this.s = s;          //s points to the current class objects
    this.start( );
  }
  public void run( ) {
    s.get( );      //retrieves the produced number and returns it to the output
  }
}
public class InterThread        //class InterThread is declared
{
  public static void main(String[ ] args)
  {
    Sort s=new Sort( );
    new Construct(s);
    new User(s);
  }
}
```

Output:

C:\nisha>javac InterThread.java

C:\nisha>java InterThread

Produce: 1

consume: 1

In this program, two threads **Construct** and **User** share the synchronized methods of the class **Sort**. At time of program execution, the **put( )** method of the **Construct** class is invoked, which increments the variable **num** by 1. After producing 1 by the Construct, the method **get( )** is invoked by the **User** class, which retrieves the produced number and returns it to the output. Thus, the **User** cannot retrieve the number without producing of it.

Caution

The three methods wait( ), notify( ), and notifyAll( ) must only be called from the synchronized methods.

Deadlock

Deadlock is a special type of error that needs to be avoided in multitasking. This mainly occurs, when two threads have interdependency on a pair of synchronized objects.

 Example: One thread enters the monitor on object A, and another thread enters the monitor on object B. If the thread in A tries to call any synchronized method on B, it
becomes blocked. But, if the thread in B tries to access any synchronized method on A, then the thread waits for a while, since it should release its own lock on B, so that the first thread can complete.

Deadlock is considered as an error, which is difficult to debug because of two reasons:

1. It occurs rarely, when two threads time-slice in the correct way.

2. It may have more than two synchronized objects and two threads.

 Example: Program that creates a deadlock situation in which two threads attempt to acquire locks on two different resources.

```
public class DLock
{
public static void main(String[ ] args)
{

    // The two resource objects for which we need to get locks are declared
    final Object res1 = "First Resource";
    final Object res2 = "Second Resource";

    //This is the first thread. It first tries to lock resource1 then resource2
    Thread thr1 = new Thread( )
    {
      public void run( )
      {
        // Lock resource1
        synchronized(res1)
        {
          System.out.println("First Thread: locked first resource");

          // Forcing deadlock to happen by pausing a bit
          try
          {
            Thread.sleep(100);
```

```
            }
        catch(InterruptedException e)
        {}

        // Wait till we get a lock on resource 2
        synchronized(res2)
        {
            System.out.println("First Thread: locked second resource");

        }
        }
    }
}


// The second thread tries to lock resource2 then resource1
    Thread thr2 = new Thread( )
    {
        public void run( )
        {
        // This thread locks resource2 without any delay
        synchronized(res2)
        {
            System.out.println("Second Thread: locked second resource");

        // Pauses like the first thread
        try
        {
            Thread.sleep(100);
        }
        catch(InterruptedException e)
        {}
```

/* The program freezes when the thread tries to lock resource 1, as thread1 has locked resource1 and shall not release it till it gets a lock on resource2. This thread holds lock on resource2 and shall not release it till it gets resource1 */

```
        synchronized(res1)
        {
            System.out.println("Second Thread: locked first resource");
```

```
          }
        }
      }
    };
```

```
    // Starts the two threads. Deadlock occurs, and the program never exits.
      thr1.start( );

      thr2.start( );
    }
  }
```

In this example,

A deadlock is created between two threads that are trying to acquire locks on two different resources. In order to avoid this type of deadlock when locking multiple resources, all the threads should acquire locks in the same order.

Task

Write a program that displays the name of the thread that executes the main method.

9.4 Suspending, Resuming and Stopping the Thread

In Java, a simple program may contain many threads. Each thread may perform different tasks. Sometimes, it becomes necessary to suspend the execution of a thread for a period of time. This can be done by using suspend( ) method of the class **Thread**. The time period needs to be specified till the thread remains suspended and then we can restart the thread by using resume( ) method of the class **Thread**. If a thread is suspended, it can be restarted. But if a thread is stopped by using stop( ) method, it cannot be restarted again.

These methods were used in the earlier systems of Java, but they are not used in the latest version because sometimes suspend( ) and stop( ) methods of the class Thread causes system failure. Therefore, run( ) method is used in the latest version of Java instead of the above mentioned methods. So, the run( ) method checks when a thread should be suspended, resumed, or stopped.

Example: Program to illustrate the concept of suspending and resuming thread.

```
class Suspenddemo     //class Suspenddemo is declared
{
public static void main(String args[])
{
Thread thrd1 = new Thread();     //new thread is created
Thread thrd2 = new Thread();  //new thread is created

try
```

```
{
Thread.sleep(1000);            //thread is made to sleep
System.out.println("Suspending: First");
thrd1.suspend( );              //current thread execution is suspended
Thread.sleep(1000);            //thread is made to sleep
System.out.println("Resuming: First");
thrd1.resume( );               //suspended thread is resumed
System.out.println("Suspending: Second");
thrd2.suspend( );              //current thread execution is suspended
Thread.sleep(1000);            //thread is made to sleep
System.out.println("Resuming: Second");
thrd2.resume( );               //suspended thread is resumed
}
catch(InterruptedException e)
{
System.out.println("The main thread is interrupted");
}
try
{
System.out.println("Applying joins");
thrd1.join( );                 //thread is joined
thrd2.join( );                 //thread is joined
}
catch(InterruptedException e)
{
System.out.println("The main thread is interrupted");
}
System.out.println("Exiting main");
}
} Output:
Suspending: First
Resuming: First
Suspending: Second
Resuming: Second
Applying joins
Exiting main
```

In this example:

1. The class Suspenddemo is declared.

2. Then, inside the main method a new instance of thread is created.

3. Then inside the try block, the first thread is put to sleep for 1000 miliseconds, after which the System.out.println statement prints the statement "Suspending: First".

4. Then, the current execution of the first thread is suspended using the method suspend( ).

5. The thread is again put to sleep, after which the System.out.println statement prints the statement "Resuming: First".

6. Now the first thread that was suspended is resumed with the method resume( ).

7. Then, the System.out.println statement prints the statement "Suspending: Second".

8. Now the current execution of the second thread is suspended using method suspend( ).

9. The second thread is put to sleep, after which the System.out.println statement prints the statement "Resuming: Second".

10. The second thread that was suspended is resumed using the method resume( ).

11. The catch method prints statement "The main thread is interrupted" if it encounters an error.

12. Then, the two threads are joined using method join( ) before which a System.out.println prints the statement "Applying joins".

13. If an error occurs while joining the threads, the catch throws an exception with the statement "The main thread is interrupted".

Finally, the System.out.println statement prints a message "Exiting main".

Caution   Do not create too many threads, as it may consume more CPU time than executing the program.

Lab Exercise

1. Write a program to create multiple threads.

2. Write a program that checks whether a given number is a prime using both the Thread class and Runnable Interface.

9.5 Summary

- A thread in running state can be resumed, suspended or blocked. Thread priorities are integer values that help to switch between threads and enable concurrent execution of code.

- The Thread class in Java is used for constructing and accessing individual thread information in a multithreaded application.

- Sometimes, it is necessary to suspend the execution of a thread. This can be done by suspend( ) method.

- Thread can be restarted by using resume( ) method. Thread can be stopped by using stop( ) method.

9.6 Keywords

Class Libraries: This is a collection of classes, which provides related facilities that can be used in the program. It is a set of classes that is stored in sets of files.

Frame: A frame is used to store data and partial results, as well as to perform dynamic linking, return values for methods, and dispatch exceptions.

Network Client: General users of the network.

Polling: This is a single queue event to determine what to do next.

9.7 Self Assessment

1. State whether the following statements are true or false:

    (a) The method Thread.SetPriorityMehtod(int) is used to set the thread priority.

    (b) Java's Thread keyword ensures that only one thread at a time is in a critical region.

    (c) The term lock refers to the access granted or approved to a particular thread to access the shared resources.

    (d) A single target object can be used only to create single threads in a program.

    (e) Invocation of the wait( ), sleep( ), or join( ) method moves the current thread to the running state.

2. Fill in the blanks:

    (a) The normal priority of a thread is _____.

    (b) The _____method of the class Thread is used to create reference of the current thread.

    (c) The _____ method is used to pause the execution of a thread for a specified period of time.

    (d) The _____method is used to check the thread (on which it is called) whether still running or not.

3. Select a suitable option in the following questions.

    (a) Which of the following methods is used to return the name of the thread group?

        (i) getName( )   (ii) getCount( )   (iii) setName( )   (iv) getMaxPriority( )

    (b) A monitor has five threads with the same priority. One of the threads is thread1. How can you notify thread1 so that it alone moves from waiting state to Ready state?

        (i) Execute notify(thread1);   (ii) Execute thread1.notify( );

        (iii) Execute notify( );   (iv) You cannot specify which thread will get notified.

9.8 Review Questions

1. "The getPriority( ) method is used to retrieve the priority of a thread". Explain this with a suitable example.

2. "Deadlock is a special type of error that needs to be avoided in multitasking". Elaborate.

3. "Thread priorities are integer values that indicate the relative priority of one thread to another". Elaborate.

4. "Java's synchronized keyword ensures that only one thread at a time is in a critical region". Comment.

5. "When a thread is alive, it indicates it is in one of its several states". Justify.

6. "A process having more than one thread is called as a multi-threaded process". Comment.

7. "The Thread class in Java (java.lang.Thread class) is used for constructing and accessing individual threads in a multi-threaded application". Elaborate.

8. "There are three ways for the threads to communicate with each other". Elaborate.

9. "An alternative way of handling synchronization is by using Synchronized Blocks". Comment.

10. "Each thread may perform different tasks. Sometimes, it becomes necessary to suspend the execution of a thread for a period of time". Comment.

11. "At certain point of time, it is necessary to terminate a thread before its task has been completed". Justify.

Answers: Self Assessment

1. (a) True   (b) False   (c) True   (d) False   (e) False

2. (a) 5   (b) currentThread( )   (c) sleep( )   (d) isAlive( )

3. (a) getName( )   (b) You cannot specify which thread will get notified

9.9 Further Readings

Books

Herbert Schildt, The Complete Reference, 7th edition, Tata McGraw Hill

D. Samanta, Object-Oriented Programming with C++ and Java

Online link

http://www.herongyang.com/Java/Deadlock-What-Is-Deadlock.html

http://java.sun.com/docs/books/jls/second_edition/html/memory.doc.html

http://tim.oreilly.com/pub/a/onjava/excerpt/jthreads3_ch6/index1.html

Unit 10: Input/Output Programming

Objectives

After studying this unit, you will be able to:

- Describe the I/O basics

- Explain the file class

- Discuss about random access files

Introduction

Any software application requires data as input. This data can be a user application, a file, or any other application. The given input is processed and the final output is produced by the software application. We know that a file is a set of records that are stored on the disk. These files can be used for storing and managing data, which is referred to as file processing. File processing comprises different operations on files, such as creating files, updating files, and so on.

In Java, all input and output operations of data using files are handled in the form of streams. A stream can be defined as a continuous series of **bytes** traveling from a source to a destination, through a communication path. The streams through the communication path can be linked with physical devices like computer screen or hard disk using I/O system in Java.

10.1 I/O Basics

Java offers a simple model for Input/Output (I/O) operations. The model for I/O operations states that data is received from a source such as a keyboard and is displayed onto the console or file. The input or the output operations are performed by writing to or reading from a stream of data. This data can be present in a file or an array, or can come from another systems' port, or can be piped from another stream.

Java provides strong and flexible assistance for I/O as it corresponds to files and networks. Most of the classes used for the I/O operations are present in the **java.io.*** package, which supports the basic I/O operations such as reading/writing data into a file. The I/O system of Java contains various classes, interfaces and methods. The I/O system of Java is cohesive and consistent. It is easy for a programmer to master the I/O system after the easy understanding of the fundamentals of this system.

10.1.1 Concept of Stream

All the applications created in Java perform I/O operations through streams. A stream is a communication path through which the data travels in a program. When a stream is sending the data, it is said to be **written**, and when the stream is receiving the data, it is said to be **read**. Data in a stream can flow only in one direction. A stream such as a file, a socket, or memory is opened by a program on an information source to get in the information and this stream is read by the program in a sequential order. Similarly, the program opens a stream to a destination by sending the written information sequentially. This is depicted in figure 10.1.

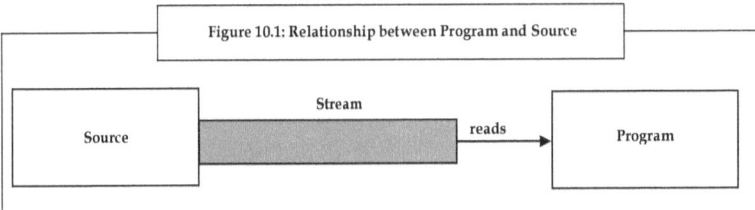

Figure 10.1: Relationship between Program and Source

As per the figure 10.1, one end of the stream must be connected to an application and the other end must be connected to the data input/output devices.

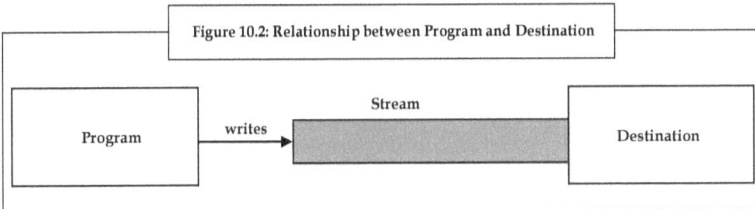

Figure 10.2: Relationship between Program and Destination

According to figure 10.1 and figure 10.2, no matter what type of information is being transferred and where the information is coming from or going to, the algorithm for reading and writing the data remains the same. Following steps are followed in these algorithms:

1. Create an object of I/O stream associated with the data source or data destination.

2. Read/write the data using the object's read( )/write( ) methods.

3. At the end, close the stream by calling the object's close( ) method.

Stream Classes

The **java.io** package comprises numerous stream classes. The stream classes in Java provide capabilities to process different types of data. Figure 10.3 depicts the categorization of stream classes.

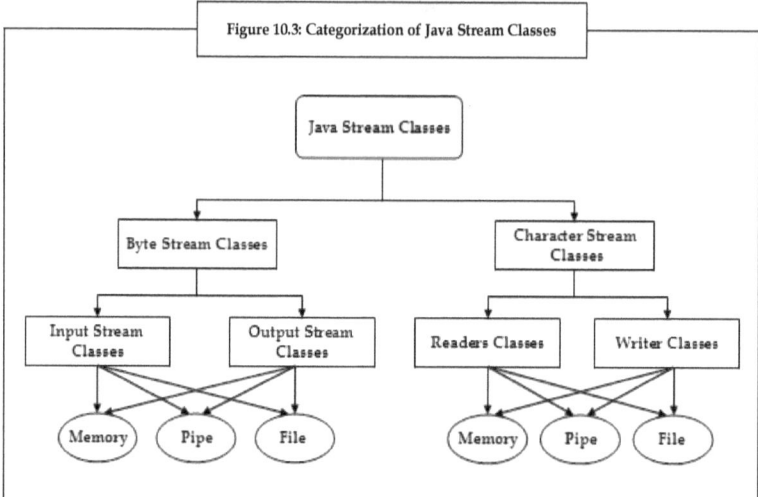

Figure 10.3: Categorization of Java Stream Classes

As depicted in figure 10.3, stream classes in Java are broadly classified into the following categories on the basis of the data type on which they operate:

1. Byte stream classes
2. Character stream classes

Byte Stream Classes

Byte stream classes are one of the categories of the Java stream classes. These stream classes handle I/O operations on bytes. There are a number of byte stream classes in **java.io** package to enable reading and writing of data as a stream of bytes. These streams act as an interface between the application and a data source or a destination. In Java, the byte stream classes are classified into two categories based on the direction of flow of data through the stream. The classes under these categories are arranged in two separate hierarchies, namely, input stream hierarchy and output stream hierarchy. These hierarchies are derived from the abstract classes as mentioned below:

1. *Input Stream Class:* It is the super class for the input stream classes.

2. *Output Stream Class:* It is the super class for the output stream classes.

These abstract classes are inherited by other subclasses to provide a variety of I/O capabilities. Each of these abstract classes comprises various subclasses such as network connections, disk files, memory buffers, and so on.

Table 10.1 depicts different byte stream classes available in Java.

Table 10.1: Different Byte Stream Classes in Java			
BufferedInputStream	BufferedOutputStream	ByteArrayInputStream	ByteArrayOutputStream
DataInputStream	DataOutputStream	FileInputStream	FileOutputStream
FilterInputStream	FilterOutputStream	InputStream	ObjectInputStream
ObjectOutputStream	OutputStream	PipedInputStream	PipedOutputStream
PrintStream	PushbackInputStream	RandomAccessFile	SequenceInputStream

To know the meaning of the byte stream classes shown in the figure 10.1, refer *"Schildt, H. (2008). The Complete Reference, 7th ed. Tata McGraw-Hill."*

Character Stream Classes

The byte stream classes provide sufficient functionalities to handle any type of input and output operations in Java. However, byte streams are not capable of handling Unicode characters. The major purpose of creating Java is to enable platform independent applications. This purpose cannot be met if the streams are unable to read/write Unicode characters. Due to this disadvantage, the character stream classes were developed. These streams can read/write characters from and to streams.

There are a number of character stream classes enabling us to read/write data as a stream of Unicode characters in **java.io** package. Character stream classes can be used to perform I/O operations on various sources or destinations of data. These streams abstract the I/O details and provide the functionalities to handle characters.

Based on the direction of flow of data in streams, the character stream classes are classified into two categories. The classes under these categories are arranged in two separate hierarchies, namely, input stream hierarchy and output stream hierarchy. These hierarchies are derived from the abstract classes as mentioned below:

1. *Reader Class:* The Reader class sets the foundation for the Reader inheritance hierarchy.

2. *Writer Class:* The Writer class sets the foundation for the Writer inheritance class.

The Reader class and the Writer class handle Unicode character streams. Table 10.2 depicts different character stream classes available in Java.

Table 10.2: Different Character Stream Classes in Java			
BufferedReader	BufferedWriter	CharArrayReader	CharArrayWriter
FileReader	FileWriter	FilterReader	FilterWriter
InputStreamReader	LineNumberReader	OutputStreamWriter	PipedReader
PipedWriter	PrintWriter	PushbackReader	Reader
StringReader	StringWriter	Writer	

To know the meaning of the character stream classes shown in table 10.2, refer *"Schildt, H. (2008). The Complete Reference, 7th ed. Tata McGraw-Hill."*

As depicted in figure 10.3, it is possible to cross-group streams depending on the type of sources they read from or write to.

 Example: Sources - memory, pipe or file.

 Notes The byte stream and character stream classes comprise specialized classes to independently handle input and output operations. The stream classes are grouped according to their functions.

10.1.2 The Predefined Streams

Java comprises many packages, and the **java.lang** package is the package that is imported by all Java programs automatically. This package consists of a class known as **System**. The **System** class encapsulates various aspects of the run-time environment.

The **System** class comprises three predefined stream variable such as **in**, **out** and **err**. Within the System class these stream variables are declared as public and static.

 Notes The stream variables are declared as public and static in a program, so that they can be used by the other parts of the program and without giving reference to a particular System object.

System.in pertains to the standard input stream such as the keyboard, which is the default input device. It is an object of **InputStream** class.

 Notes The System.in is generally not used, as data is usually passed to a command line Java application through command line arguments. The data can also be passed through configuration files.

System.out pertains to the standard output stream. It is an object of **PrintStream** class. By default, the **System.out** outputs the data written to the console. It is generally used to print a program's debug statements.

System.err pertains to the standard error stream. It is an object of the **PrintStream** class. It is usually used to output error tasks.

 *Did you know?* Eclipse, which is an IDE for running Java programs, displays the output to **System.err** in red text to help us understand easily that there is an error in the program being compiled.

 Notes System.out and System.err are mainly used to read characters and write characters from and to the console. Even though they operate on characters, they fall under byte streams. However, it is also possible to wrap these within the character based streams if required.

10.2 The File Class

A **File** class is an object used to create files, access files and directory objects, and manipulate the information of a file.

 Example: The **File** class can be used to check whether a file is hidden or not, list the directory's content, set the date of a file or set a file to read only.

As the **File** class is abstract, it represents the actual files and directory pathnames. Hence, the **File** class affords a computer independent interface for the underlying operating system.

The **File** class is also used to encapsulate information about a file. Thus, the **File** class helps us to control the file system and as such is considered as the most important class in the I/O system of the Java language.

Facts About File Class

Notes

1. The File class is present in the **java.io** package.

2. The **java.io.File** class is used to access and manipulate files and directories.

Did you know? The input and output to files is not provided by the **File** class. The **File** class just

provides an identifier/path of the files and directories.

The **File** class uses the file-naming conventions of the host operating system. The constructors of the **File** class take file and directory names as well as file paths. The constructor creates a new instance of the **File** class. After creating an instance of the file, the methods available for the **File** class can be used to:

1. Create files.

2. Delete and rename files.

3. Delete, rename and list directories.

Some constructors are used to create an object of the **File** class. These constructors are:

1. *File(String DirectoryPath):* This constructor creates a new instance of the **File** class, by specifying the path name of the file as a string parameter. This parameter includes a complete path name or a path name relative to the current working directory.

2. *File(String DirectoryPath, String Filename):* This constructor creates a new instance of a **File** class, by specifying the path name as the first parameter, and file name as the second parameter.

3. *File(String DirObj, String Filename):* This constructor creates a new instance of a **File** class in a specified directory. Here, the directory is specified as the first parameter, and the filename is specified as the second parameter. File object will be created in the current working directory only if the first parameter is **NULL**.

Example: File fileObj;

// Constructor with the file name provided as input

fileObj = new File("JavaBooks");

/* Constructor with the parent directory's path and the file name provided as inputs */

fileObj = new File("\\", "JavaBooks");

/* Constructor with the file object and file name provided as inputs. The file object represents the path to the parent directory of the file. */

File dirObj = new File("\\");

fileObj = new File(dirObj, "JavaBooks");

Note: In these examples, it is assumed that the file **JavaBooks** is present under the current directory.

The **File** class defines a wide range of methods to manage files and directories. Table 10.3 shows some of the most commonly used methods in the **File** class.

Table 10.3: Methods in File Class	
Methods	**Description**
Boolean isFileQ	Returns Boolean value, specifying whether the File object being invoked represents a File or a directory.
String getName( )	Returns the name of the File object that is being invoked, eliminating the path name.
Boolean isHidden( )	Returns Boolean value **true**, indicating that invoked File is hidden.
String getPathQ	Returns the path name of the File object that is being invoked.
Boolean canWrite( )	The application can write to the file, if the return value is true.
Boolean canRead( )	The application can read from the file, if the return value is true.
Boolean delete	Deletes the file or directory being invoked. The directory must be empty before it is deleted.
BooleanrenameTo(File directoryPath)	Used to rename the File object being invoked. The new pathname is specified to the constructor as a parameter.
longlength( )	Used to return the size of the file in bytes.
String[ ]list(FileFilter option)	Returns the file lists as an array of String objects by specifying the filter option as a parameter to the method.
File[ ]list(FileFilter option)	Returns the Files list in the Directory object being invoked. These files in the directory are returned as array of File objects.
Boolean mkdir( )	Creates a new directory if the path name of the new directory has already been specified in the File object.

The methods given in Table 10.3 are useful in creating, deleting and renaming files and directories in a file class.

10.2.1 Creating a File

The creation and utilization of a disk file involve various decisions to be made, such as an appropriate naming for the file, data type to be used, purpose of creating the file and method used in the creation of file.

A filename refers to a unique string of characters that facilitates easy identification of files on the disk. A filename consists of two parts that includes the primary name and a period with an extension.

   Example: book.txt

In this example, **book** is the primary name and **txt** is the extension.

It is important to decide the data type to be used and the type of the file stream classes that have to be used to handle the data.

   Example: Decision needs to be made if data to be handled should be of type character, byte or a non-primitive type.

The purpose of creating a file must also be decided before using the file.

   Example: The purpose of creating the file could be any, such as read only, write only or both.

In order to use a file, first it has to be opened. This is accomplished by creating a file stream and then linking it to the filename. The **Reader/InputStream** class or **Writer/OutputStream** class is created to define a file stream. The **Reader/InputStream** class is used to read data. The **Writer/OutputStream** class is used to write data. Table 10.4 and table 10.5 display the common stream classes that are used for input or output operations of bytes and characters respectively.

Table 10.4: Stream Classes Used for I/O Operations of Bytes

	Read	Write
File	FileInputStream	FileOutputStream
Memory	ByteArrayInputStream	ByteArrayOutputStream
Pipe	PipedInputStream	PipedOutputStream

Table 10.5: Stream Classes Used for I/O Operations of Characters

	Read	Write
File	FileReader	FileWriter
Memory	CharArrayReader	CharArrayWriter
Pipe	PipedReader	PipedWriter

Notes

To assign a filename to the file stream objects, we can use the constructors of the stream classes.

The file stream object can be instantiated in two ways. One way is by providing the name of the file **directly** to the constructor as a literal string or variable. The other is by providing the name of the file **indirectly** to the constructor by specifying a file object that has been assigned a filename. The following code segment depicts the use of direct and indirect method of initializing the file stream object.

 Example: Sample code to illustrate the direct method of initializing the file stream object, where the filename is specified directly inside the constructor.

```
FileInputStream fileInpStr;
try
{
    fileInptStr = new FileInputStream("BookDetails.txt");
    ...
}
catch(IOException e)
...
...
```

In this example,

1. First, a file stream object **fileInpStr** is declared.

2. Then, in the **try** block, the file stream object is initialized using the direct method. That is, a filename is assigned to the file stream object directly.

3. The **catch** block then takes care of the IOExceptions if any.

 Example: Sample code to illustrate the indirect method of initializing the file stream object, where the file object is initialized with the filename.

```
File fileObj;
fileObj = new File("BookDetails.txt");

FileInputStream fileInpStr;
try
{
    fileInptStr = new FileInputStream(fileObj);
    ...
}
```

catch(IOException e)

…

…

In this example,

1. First, a file object **fileObj** is declared.

2. Then, the filename **BookDetails.txt** is assigned to the file object.

3. After this, a file stream object **fileInpStr** is declared.

4. In the **try** block, the value of the file object is assigned to the file stream object.

5. The catch block then takes care of the IOExceptions if any.

In some programs, the **File.createNewFile( )** method is used to create a new file. If the file is created, this method returns a Boolean value **true**, otherwise it returns **false**.

 Example: Program to create a file.

```
import java.io.*;
class CreateNewFile
{
  public static void main(String args[ ])
  {
   try
    {
      File fileObj;
      fileObj = new File("BookDetails.txt");
      if(!fileObj.exists( ))
      {
         fileObj.createNewFile( );
         System.out.println("A new file \"BookDetails.txt\" has been
         created in the current directory" );
      }
    }
    catch (Exception e)
    {
      System.err.println("Error: " + e.getMessage( ));
    }
  }
}
```

Output:

A new file "BookDetails.txt" has been created in the current directory.

In this example,

1. First, the **java.io** package is imported as it supports the I/O functions.

2. Then, a class **CreateNewFile** is created.

3. In the class CreateNewFile,

    (a) The **main( )** method of the class is called.

    (b) In the **main( )** method,

        (i) The **try-catch** mechanism is used. In the **try** block, a File object **fileObj** is declared. Then, a new instance of the File class is created and is assigned the file name **BookDetails.txt**. The **if** statement is then used to find if the file name **BookDetails** already exists in the current directory. If the file does not exist, then a new file is created using the **createNewFile( )** method. Then, the **A new file \"BookDetails.txt\" has been created in** the current directory statement is printed on the screen.

        (ii) After the **try** block, **catch** block is used to handle exceptions and display the error messages if any.

Note: In this program, if the file name that is specified exists, the **createNewFile( )** method returns **false**. If the file name does not exist, the method returns **true** and creates the specified file in the current directory.

Task

Write a program to create a file and display the message "File Created" when a new file is created. In case the file already exists, the program should display the message "File Already Exists".

10.2.2 Reading and Writing Characters

The Java input/output classes facilitate to read and write characters from different sources.

The **Reader** and **Writer** class comprises various subclasses. These sub classes implement streams that can handle characters. The sub classes used to handle characters in files are **FileReader** and **FileWriter**. FileReader is used for reading characters from the file and FileWriter is used for writing characters to the file.

In the following program, these two file stream classes are used to copy the contents of a file called
source.txt into a file named **destination.txt**.

  Example: Program to copy character from one file to another file.

```
import java.io.*;

class ReadWriteCharacters
{
    public static void main(String args[ ])
```

```
{
    public static void main(String args[ ])
    {
        File sourceFile = new File("source.txt");
        File destFile = new File("destination.txt");
        FileReader fileinpstr = null;
        FileWriter fileoutstr = null;
        try
        {
            fileinpstr = new FileReader(sourceFile);
            fileoutstr = new FileWriter(destFile);
            int cha;
            while((cha = fileinpstr.read( )) != -1)
            {
                fileoutstr.write(cha);
            }
        }
        catch(IOException e)
        {
            System.out.println(e);
            System.exit(-1);
        }
        finally
        {
            try
            {
                fileinpstr.close( );
                fileoutstr.close( );
            }
            catch(IOException e)
            { }
        }
    }
}
```

Output:

Content in "source.txt" file is copied to "destination.txt" file.

In this example,

1. First, the **java.io** package is imported, as it supports the I/O functions.

2. Then, a class **ReadWriteCharacters** is created.

3. In the **ReadWriteCharacters** class,

    (a) The **main( )** method of the class is called.

    (b) In the **main( )** method,

        (i) Two File objects, **sourceFile** and **destFile** are declared and assigned the file names **source.txt** and **destination.txt**, respectively.

        (ii) Then, two file stream objects **fileinpstr** and **fileoutstr** are created and assigned the value **null**.

        (iii) Two **try** and **catch** blocks are then used.

        (iv) In the first **try** block, the two file stream objects are connected to the named files using the statements:

            fileinpstr = new FileReader(sourceFile);

            fileoutstr = new FileWriter(destFile);

            (This connects sourceFile to FileReader stream fileinpstr and also destFile to FileWriter stream fileoutstr.)

            Then, an integer object **cha** is declared. After this, the **while** statement is used to check the **((cha = fileinpstr.read( )) != -1)** condition. This condition means that the reading of characters from source file should happen until there is no character to be read in the source file. Here, the character **-1** indicates the end of the file. If this condition is true, the value of **cha** is written to **fileoutstr**. This will be done until the condition is true.

        (v) Then, in the **catch** block, the value of e is printed on the screen. Also, the **exit( )** method is used to exit the **System** class.

        (vi) After the **catch** block, the **finally** block is used. The **finally** block is executed when the **try-catch** block exits. In the **finally** block, a **try** block is used, wherein **close( )** method is used to close both the files created for reading and writing, that is, **fileinpstr** and **fileoutstr**. The **catch** block is then used to handle exceptions if any.

10.2.3 Reading and Writing Bytes

We know that the **FileReader** and **FileWriter** classes are used to read and write 16 bit characters. However, many file systems use 8 bit bytes. The I/O system of Java provides numerous classes to handle 8 bit bytes. The **FileInputStream** and **FileOutputStream** are the most frequently used classes to handle bytes.

Notes   The FileInputStream and FileOutputStream classes can be used in place of the FileReader and FileWriter classes.

In some programs, the **FileOutputStream** class is used to write bytes to a file.

 Example: Program to write bytes to a file.

import java.io.*;

```
class AddBytes
{
    public static void main(String args[ ])
    {
        byte progNames[ ] = {'J', 'a', 'v', 'a', '\n', 'D', 'o', 't', 'N', 'e', 't', '\n', 'C', '\n',
'C++', '\n', 'C#', '\n'};
        FileOutputStream fOutStr = null;
        try
        {
            foutStr = new FileOutputStream("progLang.txt");
            foutStr.write(progNames);
            foutStr.close( );
        }
        catch(IOException e)
        {
            System.out.println(e);
            System.exit(-1);
        }
    }

}
```

Output:

Java

DotNet

C

C++

In this example,

1. First, the **java.io** package is imported as it supports the I/O functions.

2. Then, a class **AddBytes** is created.

3. In the **AddBytes** class,

    (a) The **main( )** method of the class is called.

    (b) In the main **method( )**,

    (i) The following piece of code is used to declare and initialize a byte array:

    byte progNames[ ] = {'J', 'a', 'v', 'a', '\n', 'D', 'o', 't', 'N', 'e', 't', '\n', 'C',

'\n', 'C++', '\n', 'C#', '\n'};

(ii) Then, an output file stream is created and assigned the value null.

(iii) After this, a **try-catch** block is used.

(iv) In the **try** block, the following piece of code is used to instantiate the object of **FileOutputStream** with the name of the file:

foutStr = new FileOutputStream("progLang.txt");

(This creates and opens the file.)

(v) The **foutStr.write(progNames)** code is used to write data to the entire byte array to file.

(vi) Then, the file that was opened for writing is closed.

(vii) The **catch** block is used to identify and display the errors if any and to exit when there are no more characters in the source file to be read.

Task Write a program to write the names of your friends into a file named "friends.txt".

The **FileInputStream** class is used to read bytes from an existing file.

  Example: Program to read bytes from the file progLang.txt.

```
import java.io.*;
class GetBytes
{
    public static void main(String args[ ])
    {
        FileInputStream finpStr = null;
        int a;
        try
        {
            finpStr = new FileInputStream(args[0]);
            while((a = finpStr.read( )) != -1)
            {
                System.out.println((char)a);
            }
            finpStr.close( );
        }
        catch(IOException e)
        {
```

```
                    System.out.println(e);
                }
            }
        }
```

Output:

Java

DotNet

C

C++

In this example,

1. First, the **java.io** package is imported, as it supports the I/O functions.

2. Then, a class **GetBytes** is created.

3. In the class **GetBytes**,

    (a) The **main( )** method of the class is called.

    (b) In the **main( )** method,

        (i) A file input stream object **finpStr** is created and assigned the value **null**.

        (ii) Then, and integer **a** is declared.

        (iii) The **try-catch** mechanism is then used. In the **try** block, the **finpStr = new FileInputStream(args[0]);** statement is used to connect **finpStr** to the required file. Then, the **while** statement is used to check the **((a = finpStr.read( )) != -1)** condition. This condition means that characters from the source file must be read until there are no characters to be read in the source file. Here, the character -1 indicates the end of the file. If this condition is true, the value of **char a** is printed on the screen. The **close( )** method is then called to close the **finpStr** file.

        (iv) After the **try** block, the **catch** block is used to catch the **IOException e** exception. In this block, the **exception e** is printed on the screen.

Notes

The name of the file from which data needs to be read has to be given in the command line argument as follows:

java GetBytes progLang.txt

NAGESH JAITAK

10.3 Random Access Files

Random access means that the data can be read from or written to random location in a file. In the **File** class, data is read from and written sequentially as continuous streams of data. The **java.io** package comprises a class known as **RandomAccessFile** that allows performing input/output operations to any location within a file. This class also provides support for permissions such as read and write, permitting files to be accessed in read-only or read-write modes.

Two methods can be used to create a random access file. These methods are:

1.   Using the pathname as a string.

   RandomAccesFile(String pathname, String mode)

In the above syntax the **RandomAccessFile** class takes two parameters, one is the **pathname** of the location where the file is stored, and the other is the **mode** that specifies the required access permissions, such as **rw** for providing read and write permission for the file, or **r** for providing only read permission for the file.

 Example: **RandomAccessFile randomFileObj = new RandomAccesFile("test.txt", "rw"); In**

this example, a new instance **randomFileObj** of the **RandomAccessFile** is created and is assigned two arguments, one is the filename **test.txt** and the other is the access permission **rw** to the file specified in the path.

2.   Using an object of the File class.

   RandomAccessFile(File name, String mode)

In the above syntax, **RandomAccessFile** class takes two parameters, one is the file object that consists of the **file name**, and the other is the **mode** that specifies the required access permissions such as **rw** for providing read and write permission for the file, or **r** for providing only read permission for the file.

 Example: **File fObj = new File("test.txt");**

   RandomAccessFile randomFileObj = new RandomAccesFile(fObj, "rw");

   In this example,

   1.   An object **fObj** of the **File** class is created using the **new** keyword, and is assigned the filename **test.txt**.

   2.   A new instance **randomFileObj** of the **RandomAccessFile** class is created with two arguments, one is the file object **fObj** that comprises the file name and the other is the read/write **(rw)** permission for the file.

The **RandomAccesFile** class offers various methods to randomly access locations inside a file. Table 10.6 provides a list of such methods along with their usage.

Table 10.6 Methods of RandomAccessFile Class

Methods	Usage
int writeBytes(String)	Writes the String i.e. specifies as bytes to a file.
void seek(long loc)	Sets the file pointer to a specified location within the file.
Byte [ ] readBytes(long)	Reads specified number of bytes from a file to a byte array.
long getFilePointer( )	Returns the current location of a file pointer.
long length( )	Returns length of a file in bytes.

As per the table 10.6, various reading and writing operations can be performed using a random access file class. In some programs, a file pointer supported by the random access files is used. This file pointer is moved to arbitrary positions in the file before reading or writing. The **seek( )** method of the **RandomAccessFile** class is used to move the file pointer.

 Example: Program to read and write using random access file.

```
import java.io.*;
class RandReadWrite
{
    public static void main(String args[ ])
    {
        RandomAccessFile raFile = null;
        try
        {
            raFile = new RandomAccessFile("random.txt", "rw");
            raFile.writeChar('Java');
            raFile.writeDouble(6.0);
            raFile.writeInt(2011);
            raFile.seek(0);
            System.out.println(raFile.readChar( ));
            System.out.println(raFile.readDouble( ));
            System.out.println(raFile.readInt( ));
            raFile.seek(2);
            System.out.println(raFile.readDouble( ));
```

```
        raFile.seek(file.length( ));
        raFile.writeBoolean(false);
        raFile.seek(4);
        System.out.println(file.readBoolean( ));
        file.close( );
    }
    catch(IOException e)
    {
        System.out.println(e);
    }
  }
}
```

Output:

Java

6.0

2011

6.0

false

In this example,

1. First, the **java.io** package is imported, as it supports the I/O functions in a program.

2. Then, a class **RandReadWrite** is created.

3. In the class **RandReadWrite**,

  (a) The **main( )** method of the class is called.

  (b) In the **main( )** method,

    (i) An object of the **RandomAccessFile class (raFile)** is created and is assigned the value **null**.

    (ii) Then, in the **try** block, the **raFile = new RandomAccessFile("random.txt", "rw");** statement is used to open the specified file **random.txt**.

    (iii) The **raFile.writeChar('Java');** statement is used to write the value of **Char** data type Java into **raFile**.

    (iv) Then, **raFile.writeDouble(6.0);** statement is used to write the values of **Double** data type **6** and **0** into **raFile**.

    (v) The **raFile.writeInt(2011);** is then used to write the value of **Int** data type 2011 into **raFile**.

    (vi) Then, the **seek( )** method of **raFile** is called with the value of **0**, which is used to is used to bring the file pointer to the beginning of the file.

(vii) The value of **Char** data type in **raFile** is then read and printed on the screen.

(viii) Then, the value of **Double** data type in **raFile** is read and printed on the screen.

(ix) The value of **Int** data type in **raFile** is then read and printed on the screen.

(x) The **seek( )** method of **raFile** class is again called but with the value of **2**, which is used to bring the file pointer to the second position of the file.

(xi) Again, the value of **Double** data type in **raFile** is read and printed on the screen.

(xii) The **seek( )** method of **raFile** is then called along with the call to the **length( )** method to find the length of the file.

(xiii) After this call, **raFile.writeBoolean(false);** statement is used write the Boolean value **false** into the **raFile**.

(xiv) The **seek( )** method of **raFile** is again called, but with the value of **4**, which is used bring the file pointer to the fourth position of the file.

(xv) Then, the Boolean value is read and printed on the screen.

(xvi) The **close( );** method is used to close the file.

4. After the **try** block, **catch** block is used to catch the **IOException e** exception. In the **catch** block, the value of exception **e** is printed on the screen.

Note: In this example, the RandomAccessFile class constructor assumes that the file that is specified as parameters exits in the current directory. Therefore, an empty file named as random.txt has to be created if you are executing the program for the first time.

Lab Exercise

1. Write a program to append data items to an existing file on your system.

2. Write a program to insert the name of any five hardware components of a computer into a .txt file by using the **FileOutputStream** class.

10.4 Summary

- The I/O operation refers to the receiving and displaying of data from or to a file.

- The classes used for the I/O operations are present in the **java.io.*** package.

- A stream refers to a communication path, using which the data travels in a program.

- The algorithm for reading and writing data includes creating an object of I/O stream, reading or writing the data using the read( ) or write( ) methods and closing the stream by calling the close( ) method.

- When a stream sends data, it is said to be **written**; when the stream receives data, it is said to be **read**.

- Stream classes are of two types, namely, byte stream classes and character stream classes. Byte stream classes handle I/O operations in bytes.

- Based on the direction of flow, byte streams classes are classified into two types, that is, input stream class and output stream class.

- Character stream classes are used to handle reading or writing of Unicode characters from and to streams.

- Based on the direction of flow, character stream classes are classified into two types, that is, reader class and output stream class.

- The three predefined stream variables present in the **System** class are **in**, **out** and **err**.

- The **File** class is used to create, access, and manipulate the files and directory objects.

- Creation of a file involves decisions to be taken regarding the appropriate naming of the file, using data types, purpose of creating the file and method used to create the file.

- The sub classes of the Reader and Writer classes are used to read and write characters into a file, respectively.

- FileReader and FileWriter classes are used to handle characters.

- FileInputStream and FileOutputStream classes are used to handle bytes.

- The RandomAccessFile class is used to perform input/output operations to any location within a file.

10.5 Keywords

Algorithm: It refers to a set of rules specifying the method to solve some problem.

Console: An output device.

IDE: Integrated Development Environment

Socket: One end point of a two way communication link between two programs that are running on a network.

Unicode: It is a 16-bit character set standard, designed to include characters appearing in most languages.

10.6 Self Assessment

1. State whether the following statements are true or false:

    (a) When a stream is receiving the data, it is said to be written, and when the stream is sending the data, it is said to be read.

    (b) The byte stream classes provide sufficient functionalities to handle any type of input and output operations in Java.

    (c) The FileInputStream class encapsulates various aspects of the run-time environment.

    (d) System.out is an object of PrintStream.

    (e) FileReader is used for reading characters from the file.

    (f) The System class is used to read bytes from an existing file.

2. Fill in the blanks:

    (a) A _____is a communication path, through which the data travels in a program.

    (b) The _____classes are those stream classes that handle I/O operations on bytes.

    (c) Most of the classes used for the I/O operations are present in the _____package.

    (d) The _____ classes can be used to perform I/O operations on various sources or destinations of data.

    (e) In the File class, data is read from and written sequentially as continuous streams of data and thus are known as _____files.

3.   Select a suitable choice in every question.

  (a)   Which of the following is the super class for the input stream classes?

   (i)   Reader class

   (ii) Writer class

   (iii) Output stream class

   (iv) Input stream class

  (b)   Which of the following is a class under the character stream class?

   (i)   PrintWriter

   (ii) PipedReader

   (iii) FileWriter

   (iv) PipedOutputStream

  (c)   Which of the following is the stream class used for I/O operations of bytes?

   (i)   CharArrayReader

   (ii) ByteArrayInputStream

   (iii) CharArrayWriter

   (iv) PipedWriter

  (d)   Which of the following classes is used to handle bytes?

   (i)   FileWriter

   (ii) StringReader

   (iii) FileOutputStream

   (iv) BufferedReader

  (e)   Which of the following classes can be used to read from or write to any location in a file?

   (i)   PrintWriter

   (ii) RandomAccessFile

   (iii) Reader

   (iv) Writer

10.7   Review Questions

1.   "Java offers I/O operations for input or output of data". Elaborate.

2.   "No matter what type of information is being transferred and where the information is coming from or going to, the algorithm for reading and writing the data remains the same." Write the steps followed in these algorithms.

3.   "All the applications created in Java perform I/O operations through the functionality of streams." Justify.

4.   "The **java.io** package comprises numerous stream classes." Elaborate.

5.   "There are a number of byte stream classes in **java.io** package to enable reading and writing of data as a stream of bytes." Justify.

6.   "Based on the direction of flow of data in streams, the character stream classes are classified into two categories." Justify.

7. "There are a number of character stream classes enabling us to read/write data as a stream of Unicode characters in java.io package." Justify.

8. "Java comprises **java.lang** package, which consists of a class known as **System**." Elaborate.

9. Analyze the usefulness of the **File** class of the **java.io** package.

10. "RandomAccessFile class takes two parameters, one is the file object and the other is the mode." Comment.

11. "Two file stream classes can be used to copy the contents of one file to another." Discuss.

12. "Random access means that the data can be read from or written to random location in a file." Justify using a program.

Answers: Self Assessment

1. (a) False

    (b) True

    (c) False

    (d) True

    (e) True

    (f) False

2. (a) Stream

    (b) Byte stream

    (c) java.io.*

    (d) Character stream

    (e) Sequential

3. (a) PipedOutputStream

    (b) PrintWriter

    (c) ByteArrayInputStream

    (d) FileOutputStream

    (e) RandomAccessFile

10.8 Further Readings

Books

Schildt. H. (2002), Java 2 The Complete Reference, 5th ed. New York: McGraw-Hill/Osborne.

Pravin Jain M. (2011), The Class of Java. India: Dorling Kindersley (India) Pvt. Ltd.

Online link

http://home.cogeco.ca/~ve3ll/jatutor9.htm

http://www.hostitwise.com/java/java_io.html

http://tutorials.jenkov.com/java-io/index.html

Unit 11: Introduction to Applets

CONTENTS
Objectives
Introduction
11.1 Fundamentals of Applets
11.2 Applet Life Cycle
11.3 Applet Tag
11.4 Running the Applet
11.5 Handling Images and Sound
11.5.1 Images
11.5.2 Sound
11.6 Summary
11.7 Keywords
11.8 Self Assessment
11.9 Review Questions
11.10 Further Readings

Objectives

After studying this unit, you will be able to:

- Describe the fundamentals of applets
- Discuss the life cycle of applets
- Describe applet tags
- Illustrate the method to run applets
- Explain the handling of images and sound

Introduction

Java is programming language that is widely used in the development of numerous software and applications, such as the creation of interactive Web pages. The creation of Web pages is related to Java applets. An applet is an Internet based Java program that can be included in an HTML page and can be downloaded on any computer. Applet is also popularly known as a programming language for the Web.

Did you know?   Netscape Communications (formerly known as Netscape Communications Corporation), a computer services company in the USA, added Java support to its popular Navigator in 1996. After this the Web changed from static Web pages to exciting Web pages with the use of applets.

11.1 Fundamentals of Applets

Applets are special Java programs that are embedded in Web pages. They can be transmitted over the Internet and automatically executed by using a Java-compatible Web browser present in the user's system.

 Example: When a user opens a Web page, the applet also gets executed automatically. The user need not install any special software in order to run an applet on the Web page.

Applets are not just used in media files or animations. They can also be used for performing calculations, displaying graphics, creating animations, playing sounds and running interactive games as application programs. Application programs are the programs that can react to user's inputs.

Did you know? Applets can access remote databases or operate on complicated data on the user's system.

With the use of applets, Java has changed the way in which the Internet users retrieve and use documents on the World Wide Web (WWW). It facilitates the users to create and use completely interactive multimedia Web documents.

Basically, an applet is a Java class defined in JDK's **java.applet** package. This package provides all facilities to control the execution of applets.

There are two methods of embedding applets into Web pages. They are:

1. *By Writing Our Own Applets and Embedding Them into Web Pages:* An applet that is developed and stored in a local system is termed as **local applet**. In case a Web page has to access a local applet, it has to merely search the directories in local system, and identify and load the specified applet. Hence, this process does not need an Internet connection.

2. *By Downloading Applets from a Remote system and Embedding Them into Web Pages:* An applet that is developed and stored on a remote system is termed as **remote applet.** If a user's system is connected to the Internet, the remote applet could be downloaded and run on the user's system. It is important for the user to know the location or address of the remote applet on the Web to access it. The location or address of the remote applet is known as Uniform Resource Locator (URL).

Notes The URL of the remote applet is specified in the applet's HTML document as a value for the attribute CODEBASE.

CODEBASE = http://www.remoteservices.com/applets

While accessing a local applet CODEBASE, attributes can be used to specify the path of the local applet on the system. However, it is not mandatory to use the CODEBASE attribute to access a local applet.

Types of Applets

Applets can be classified into two types. These types are:

1. *Based on the Applet Class:* The first type of applets is based directly on the **Applet** class. These applets utilize the Abstract Window Toolkit (AWT) to provide Graphical User Interface (GUI).

Notes   The java.applet package contains the Applet class.

2.   ***Based on Swing Class JApplet:*** The second type of applet is based on the Swing class **JApplet.**
These applets utilize Swing classes to provide GUI. When compared to AWT, Swing often provides an easy-to-use user interface. Therefore, the Swing-based applets are very popular. However, people still prefer to use AWT-based applets to create simple user interfaces.

Notes JApplets inherits all the properties of the Applet class.

Applet Architecture

Applets share a common architecture and have the same life cycle, whether they are based on Applet or on JApplet, and resemble a window-based GUI program. Thus, it can be said that they are not organized as a console-based program. The execution of an applet does not begin with the **main( )** method. However, some applets have the **main( )** method.

Notes   Console-based programs refer to command line programs that do not require a GUI to run.

The execution of an applet is controlled by the life cycle methods. **System.out.println( )** is not used to get the output for an applet's window and **readLine( )** method is not used for input operations. Instead, numerous controls offered by the Swing or AWT components are used to handle user interactions. We can write an output directly to an applet window using methods such as **drawString( )** instead of **println( ).**

Java applets are event-driven. This means that an applet waits for an event to occur.

   Example: Selecting an item from a list or clicking of the mouse button by the user.

The applet gets notification from the runtime system about the event by calling an event handler provided by the applet. After this, the applets have to take appropriate actions in response to the events and return the control to the runtime system.

11.2 Applet Life Cycle

As mentioned earlier, applets are executed within a Web browser or in an applet window. When an applet is executed, it undergoes four phases; these four phases are known as the life cycle of an applet. The four phases present in the life cycle of an applet are:

1.   Initialize

2.   Running

3.   Stop

4.   Dead

Figure 11.1 depicts the life cycle of an applet and the different methods used in the phases of this life cycle.

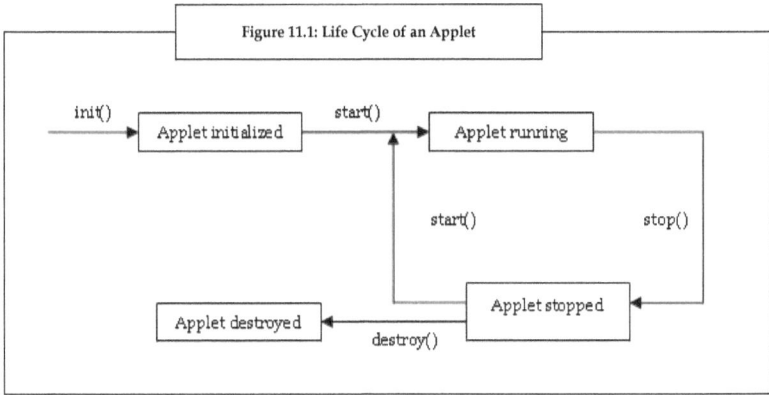

Figure 11.1: Life Cycle of an Applet

As depicted in the figure 11.1, the phases in the life cycle of an applet are:

1. *Initialize:* When an applet is first loaded, it enters the initialization phase. The **init( )** method of the Applet class is used to initialize the applet code. This method is called only the first time an applet is loaded in the computer's memory. The **init( )** method is used to initialize variables, load images or fonts, and add components such as buttons and text fields to the applet.

2. *Running:* When an applet calls the **start( )** method of an Applet class, it enters the running phase. This happens automatically after the applet is initialized and every time it receives focus.

    Example: When a user leaves a Web page containing an applet and returns back to the same page after some time, the applet starts running again. That is, the applet will return back to the running phase.

    Hence, the **start( )** method is used to restart a process every time a user visits a Web page. The **start( )** method can be called any number of times unlike the **init( )** method that is called only once.

3. *Stop:* When an applet is stopped from running, it enters the stopped phase. This happens automatically when the user leaves a Web page containing an applet. It is also possible to explicitly call the **stop( )** method.

    Example: If a thread is used to run an applet, then the stop( ) method has to be used to suspend the thread.

    The **stop( )** method is used to reset variables and stop the threads that are running.

4. *Destroy:* When an applet is removed from the computer's memory, it enters the destroy phase. This happens automatically when the **destroy( )** method is invoked or when a user quits the Web browser. The **destroy( )** method can be used to perform clean-up operations such as closing a file. The destroy phase also occurs only once like the initialize phase.

Notes   It is not always necessary for the applet to override the destroy( ) method, since the stop( ) method performs all tasks required to shut down the applet's execution. However, the destroy( ) method has to be used for applets that need to release additional resources.

It is not an obligation to use any or all of the above methods of an applet. The Java environment calls the above methods automatically. Therefore, we must declare these methods as public.

Caution   We must not add parameters to any of the above methods, as these methods do not accept parameters.

11.3 Applet Tag

In Java, two types of tags can be used, that is, a parameter tag and an applet tag. A parameter tag is a tag, which names a parameter that needs to be run by the Java applet, along with the parameter's value. An applet tag is the tag used to add Java applets in the HTML documents. The applet tag is written within the BODY tag of an HTML document.

Syntax of the Java Applet Tag:

<APPLET

    CODE = "name of the .class file"

    CODEBASE = "path of the .class file"

    HEIGHT = "maximum height of the applet, in pixels"

    WIDTH = "maximum width of the applet, in pixels"

    VSPACE = "vertical space between the applet and the rest of the HTML, in pixels"

    HSPACE = "horizontal space between the applet and the rest of the HTML, in pixels"

    ALIGN = "alignment of the applet with respect to the rest of the Web page"

    ALT = "alternate text to be displayed if the browser does not support applets">

  <PARAM NAME = "parameter_name" VALUE = "value_of_parameter">

  <PARAM NAME = "parameter_name" VALUE = "value_of_parameter">

...

</APPLET>

In this syntax,

1. <APPLET> </APPLET> is the applet tag.

2. CODE, CODEBASE, HEIGHT, WIDTH, VSPACE, HSPACE, ALIGN, ALT included within the applet tag are known as **attributes** of the applet tag. These are the most commonly used attributes of the APPLET tag.

3. Parameters are supplied to the applet through the <PARAM> tag. The PARAM tag must be written between <APPLET> and </APPLET>.

Example: The HTML code for placing the applet named as **DisplayApp** on a Web page:

```
<HTML>
 <HEAD>
 </HEAD>
   <BODY>
    <APPLET
       CODE = "DisplayApp.class"
       HEIGHT = 300
       WIDTH = 500
    >
    </APPLET>
    </BODY>
</HTML>
```

In this code,

1. *<HTML> </HTML>*: This HTML tag indicates the start and end of a HTML file.

2. *<HEAD> </HEAD>*: This tag may include details about the Web page, that is, it can contain the <TITLE> tag used to add text that has to be displayed on the title bar of the browser.

3. *<BODY> </BODY>*: This tag includes the main text of the Web page. The
    <APPLET> tag has to be declared within this <BODY> tag.

A pair of applet tags is included in the body section of the HTML (Hyper Text Mark-up Language) code. These applet tags specify the name of the applet to be loaded using the CODE attribute. The CODE attribute informs the browser to load the **DisplayApp.class** applet on the Web page. The applet tag also informs the browser about the display area for the applet output using the HEIGHT and WIDTH attribute. Here, the height and width of the applet is assigned a value of
300 and 500 pixels respectively.

Notes
It is essential to store this HTML file in the same directory as the DisplayApp.class applet.

11.4 Running the Applet

It is necessary to create an executable applet before we learn to run it.

Creation of an executable applet refers to the creation of a **.class** file. The **.class** file can be obtained by compiling the source code of the applet program. The method of compiling an applet is same as compiling an application in Java. Hence, the Java compiler is used to compile the applet.

 Example: Program to create and compile a simple Java applet code.

```
import java.awt.*;

import java.applet.*;

public class DisplayApp extends Applet

{
```

```
public void paint (Graphics a)
{
    a.drawString("My First Applet Program", 15, 100);
}
}
```

Output:

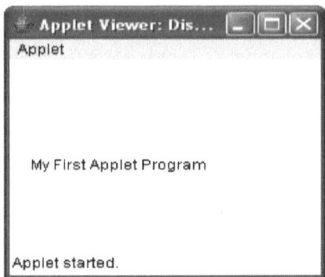

In this example,

1. First, **import.java.awt.*** package is imported for using the Graphics class in the program. (We need the Graphics class, as it contains all the methods required to perform the output operations.)

2. Then, **import.java.applet.*** package is imported for using the Applet class. (We need the Applet class, as it contains methods that provide life and behavior to applets such as **init( ), start( ), stop( )** and **destroy( )**).

3. After importing the necessary packages, a class **DisplayApp** is created that extends the **Applet** class. This is the main class for the applet.

4. In the class **DisplayApp,**

    (a) The **paint( )** method of the **Applet** class is called to display the output of the applet code. This **paint( )** method has taken a **Graphics** object **a** as an argument.

    (b) In the **paint( )** method, the **a.drawString("My First Applet Program", 15, 100)** statement is used to draw the string **My First Applet Program** at the position 15, 100 (pixels), when the applet is executed.

To execute the applet code, first go to the command prompt and move to the directory that contains the applet code and type the **javac DisplayApp.java** command. Then, the compiled output file **DisplayApp.class** is placed in the same directory as the source.

In order to run an applet, either a Java-enabled Web browser such as Internet Explorer, Netscape Navigator, or an appletviewer is required.

If the applet is run using a Web browser, open the browser window and insert the path where the HTML file is stored in the system.

If the applet is run using an appletviewer, type the following command in the command prompt:

appletviewer DisplayApp.html

Notes

| | 1. | If a Java enabled browser is used to run the applet code, then the entire Web page containing the applet is seen. |
| | 2. | If the appletviewer tool is used to run the applet code, then only the applet output is seen. |

11.5 Handling Images and Sounds

Applets can be used to handle images and sounds. This means that we can either display an image or play sounds using applets.

11.5.1 Images

Images are generally used to build a professional-looking user interface. They are important components of Web design. Images can be displayed using GIF and/or JPEG format in Java.

The **java.awt** package consists of all the classes required to manipulate images. The images are objects of the **Image** class. To add an image to an applet, one needs to import the **java.awt.images** package.

Did you know?   CompuServe, which was the first major commercial online service provider in the USA, created the GIF image format in the year 1987 to make it possible to view online images. However, GIF images supported only up to 256 colors in each image. To overcome this limitation, a group of photographic experts created the JPEG format. This format stores full-color-spectrum and continuous-tone images.

There are three operations to be considered while working with images. These operations are:

1. Creating an image object
2. Loading an image
3. Displaying an image

Creating an Image Object

To create an image object, the **createImage( )** method of the **Component** class in the **java.awt** package is used.

Syntax of Creating an Object

Image createImage(int width, int height)

In this syntax, **Image** is the class wherein **createImage** method is used to create an off-screen drawable image with specified **width** and **height**. These parameters, that is, width and height are of **integer** type.

   Example: Canvas cnv = new Canvas( );
Image i = cnv.createImage(300, 200);

In this example, first an instance of Canvas class is created using the **new** keyword. Then, an image object **i** is created by calling the **createImage( )** method with width **300** and height **200**.

Loading an Image

It is also possible to load an image using the **getImage( )** method of the Applet class. An image can be loaded in two ways:

1. ***Image getImage(URL u):*** This method returns an Image object that encapsulates the image that is present in the location specified by the URL (Uniform Resource Locator).

2.   *Image getImage(URL u, String imgNm):* This method returns an Image object that encapsulates the image that is present in the location specified by the URL and also has the name specified by the **imgNm**.

Displaying an Image

After loading an image, it can be displayed by using the **drawImage( )** method of the Graphics class.

 Example: Program to load and display an image named **javacup.gif**.

```
import java.awt.*;
import java.applet.*;
public class DisplayImage extends Applet
{
    Image i;
    public void init( )
    {
        i = getImage(getDocumentBase( ), "javacup.gif");
    }
    public void paint(Graphics gra)
    {
        gra.drawImage(i, 0, 0, this);
    } }
```

Output:

In this example,

1.   First, the two packages **java.awt.*** and **java.applet.*** are imported using the **import** keyword.

2.   Then, a class **DisplayImage** is created that extends the **Applet** class.

3.   In the **DisplayImage** class,

     (a)   An Image object i is declared.

     (b)   Then, the **init( )** method is called.

(c) In the **init( )** method, **i = getImage(getDocumentBase( ), "javacup.gif");** statement is used. In this statement, **getImage( )** method is called with two parameters, that is, the **getCodeBase( )** and the filename of the image. The output of the **getImage( )** method is assigned to **i.** Then, the **paint( )** method is called to draw the graphics object **gra** of the applet in the drawing area.

(d) In the **paint( )** method, **gra.drawImage(i, 0, 0, this);** statement is used to draw the graphics object **gra** with the image input **i,** the top left location **0** and **0.**

11.5.2 Sound

Apart from images, sounds can also be used to make Web pages more interesting and entertaining. Similar to images, sound files can also be added to an applet.

Playing a Sound File

Using a sound file in an applet is quite simple.

 Example: Program to play a sound file when an applet is downloaded.

```
import java.applet.*;
public class AppletSound extends Applet
{
    public void init( )
    {
        super.init( );
        resize(0,0);
        AudioClip song = getAudioClip(getDocumentBase( ), "song.au");
        song.play( );
    }
}
```

In this example,

1. First, the **java.applet.*** package is imported.

2. Then, a class **AppletSound** is created, which extends **Applet** class.

3. In the **AppletSound** class, the **init( )** method is called.

4. In the **init( )** method,

(a) First, the **super.init( );** statement is used to call the **init( )** method on the **super** class.

(b) Then, the applet size is set to zero pixels to hide the applet on the Web page.

(c) The **AudioClip song = getAudioClip(getDocumentBase( ), "song.au");** statement is then used to load the sound file into the AudioClip object **song.**

(d) Then, **song.play( );** statement is used to play the sound file **song** that is present under the AudioClip class.

Note: While compiling the above code, the sound file should be stored in the same directory as the class file. In case the sound file is stored in a different directory, the complete path must be specified.

Lab Exercise　　　Write an applet program to display as well as play a sound track.

11.6　Summary

* Applets are Java programs embedded in Web pages and can be transferred over the Internet and executed automatically using the applet viewer.

* Applet is a Java class that is defined in the JDK's **java.applet** package.

* We can embed applets within Web pages by either writing our own applets and embedding them into Web pages or by downloading an applet from a remote system and embedding them into Web pages.

* Applets that are developed and stored in local systems are known as local applets and applets that are developed by another person and stored on a remote system are known as remote applets.

* There are two types of applets. One is based on the Applet class and the other is based on the Swing class JApplet. The execution of an applet is controlled by the life cycle of applets. Applet tags are used to include Java applets.

* The four phases of the applet life cycle are initialize, run, stop and dead.

* To run an applet program a .class file of the applet program has to be created. Applet programs can be run either by using an HTML file or an appletviewer.

* Images can be loaded and displayed using applets in a Web page. Sound can also be played using applets.

11.7　Key words

Canvas Class: Provides a display area upon which graphical output can be produced, or a specialized user interface component can be produced.

GIF: Graphic Interchange Format. *HTML:*

Hyper Text Mark-up Language. *JPEG:*

Joint Photographic Experts Group.

Tags: A sequence of characters in a mark-up language to provide information specific to formatting or display of text on a Web browser.

11.8　Self Assessment

1.　State whether the following statements are true or false:

(a)　The concept of applet is used by the programmers for creating Graphical User Interface (GUI) objects, such as scroll bars, buttons, and windows.

(b)　An applet that is developed and stored in a stand alone system is termed as local applet.

(c)　Applets that are based on the swing class JApplet, utilize the Abstract Window Toolkit (AWT) to provide Graphical User Interface (GUI).

(d)　When an applet calls the **start( )** method of an Applet class, it enters the running phase.

2.　Fill in the blanks:

(a)　The location or address of the remote applet is known as _____.

(b) The _____attribute can be used to specify the path of the local applet on the system.

(c) When an applet is first loaded it enters the _____phase.

(d) Creation of an executable applet refers to the creation of a _____file of the applet that is got by the compilation of the source code of the applet.

3. Select a suitable choice in every question.

(a) When an applet is removed from the computer's memory it enters which of the following phases?

    (i) Destroy   (ii) Initialize   (iii) Run   (iv) Stop

(b) Which of the following methods are used to display the output of the applet code?

    (i) drawString( )   (ii) paint( )   (iii) drawImage( )   (iv) getImage( )

(c) Which of the following methods will return an Image object that encapsulates the image that is present in the location specified by the URL?

    (i) Image getImage(URL u, String imgNm)   (ii) getDocumentBase( )

    (iii) drawImage( )   (iv) Image getImage(URL u)

11.9 Review Questions

1. "Applets are of two types". Discuss.

2. Analyse the architecture of Java applets.

3. "When an applet is executed, it undergoes four phases ". Justify.

4. "An applet tag is used to add Java applets". Elaborate.

5. Assume you have created a simple applet program named FirstApplet. What is the procedure to be followed to run the applet using Internet Explorer?

6. "There are three operations to be considered while working with images". Elaborate.

Answers: Self Assessment

1. (a) False   (b) True   (c) False   (d) True

2. (a) Uniform Resource Locator (URL)   (b) CODEBASE (c) Initialization   (d) .class

3. (a) Destroy   (b) paint( )   (c) Image getImage(URL u)

11.10 Further Readings

Books

Balagurusamy E. Programming with Java_A Primer 3e. New Delhi

Schildt. H. Java 2 The Complete Reference, 5th ed. New York: McGraw-Hill/Osborne.

Online link

http://www.dgp.toronto.edu/~mjmcguff/learn/java/

http://www.realapplets.com/tutorial/index.html

Unit 12: Abstract Window Toolkit

Objectives

After studying this unit, you will be able to:

- Discuss the fundamentals of AWT controls

- Explain AWT layout managers

- Illustrate the use of various layout managers

Introduction

One of the important components of Java is AWT (Abstract Window Toolkit). AWT is a single window interface, which can work on multiple platforms. The concept of AWT is used by the programmers to create Graphical User Interface (GUI) objects such as scroll bars, buttons, and windows. AWT is a part of Java Foundation Class (JFC), which is a broad set of GUI class libraries. The GUI class libraries enable the user to develop the user interface part of an application program.

12.1 AWT Control Fundamentals

AWT refers to a class library provided by the Java programming language. AWT comprises many graphical widgets that can be used in the display area with a layout manager.

AWT is platform-independent like the Java programming language. AWT offers a common set of tools to facilitate programmers to design the GUIs, which work the same way on different platforms. The user interface elements offered by the AWT are implemented by using the platform's native GUI toolkit, which helps to preserve the look and feel of each platform.

A GUI is built using graphical elements known as components. **Component** class is the super class of all GUI objects such as buttons, scrollbars, and text fields. In the AWT, all the user interface components are instances of the Component class or one of its subtypes. AWT facilitates user interaction with the program.

Components are not stand alone. They are placed within containers. Containers control the components of the layout. Containers can also be placed inside other containers as they themselves can be considered as components. All containers in AWT are instances of the class Container or one of its subtypes.

Did you know?   Components are known as Java's building blocks for creating GUIs.

Characteristics of AWT are:

1.   It consists of a set of user interface components.

2.   It is a robust event-handling model.

3.   It has tools for graphics and imaging, which includes shape, color, and font classes.

4.   It comprises layout managers, for flexible window layouts that are independent of a particular window size or screen resolution.

5.   It includes data transfer classes for cut-and-paste through the native platform clipboard.

Notes   The java.awt.* is an AWT package that is used to develop the user interface objects such as buttons, radio buttons, list box, menus, and so on. This package comprises various classes and interfaces helpful in the development of GUI applications.

Adding and Removing Controls

To include a control in a window, the control needs to be first added to the window. This is done by creating an instance of the desired control and then adding it to a window by calling the **add( )** method. The **add( )** method is defined by the **Container**.

Example: Component add(Component *compObj*);

In this example, **compObj** is an instance of the control that needs to be added. Hence, a reference to **compObj** is returned.

Once a control has been added, it is visible whenever its parent window is displayed.

Responding to Controls

There may be times when a control from a window needs to be removed. This is generally the case when the control is no longer in use. Under such circumstances, we call the **remove( )** method. The **remove( )** method too is defined by the **Container**. The syntax for this is:

void remove(Component *Obj*)

In this syntax, **Obj** is the reference to the control that needs to be removed. All controls can be removed by calling **removeAll( )** method.

All controls, when accessed by the user, generate events. This means, the program implements the appropriate interface and then registers an event listener for each control that needs to be monitored.

Example: When a user clicks on a push button, an event that identifies the push button is sent.

AWT has a collection of basic user interface components and it allows the user to create their own components. Modern user interfaces are built on the concept of components. The various components of AWT are discussed in the following sub-sections.

12.1.1 Labels

A label is a simple component of Java AWT. Labels are used to show the text or string on a user's application and do not perform any type of action.

The following piece of code illustrates how to create a label control.

Label label_name = new Label ("This is the label text.");

The following piece of code creates a label control and aligns it in the center.

Label label_name = new Label ("This is the label text." Label.CENTER);

The alignment of label can be left, right or centered. In the syntax provided above, center justification of the label is represented using the name Label.CENTER.

12.1.2 Buttons

A button is used to trigger actions and other events required for user's application.

The following piece of code illustrates how to create a button control.

Button button_name = new Button("This is the label of the button");

Buttons are added to its container using add(button_name) method.

12.1.3 Check Boxes

This component of Java AWT allows users to create check boxes in their applications.

The syntax for defining a Checkbox is given below:

CheckBox checkbox_name = new Checkbox("Optional check box 1", false);

In the syntax given above, the code is used to construct an unchecked checkbox. This is done by passing a Boolean value argument, either true or false, with the checkbox label.

12.1.4 Choice Lists

A choice list is used to create a pop-up menu from which a user can choose only a single item. The current item that is selected by the user is displayed as the title of the choice menu.

An instance of the Choice class must be created to work with choice list.

Choice theoptions = new Choice( );

After creating the choice, the addItem method is used to add new entries.

    theoptions.addItem ("White");

    theoptions.addItem ("Black");

The user can change the currently selected item by name or index.

    theoption.select ("Red");

    theoption.select (0);

The position and string name of the selected item are returned by the **getSelectedIndex( )** and **getSelectedItem( )** methods respectively.

Whenever a selection is made, the Choice class calls the action method irrespective of whether it is a new selection or not. The name of the selected item is contained in the parameter **whatAction.**

12.1.5 Lists

A List box permits a user to create a scrolling list of values, from which a user can either select a single item or multiple items.

If the user wants to create a list that does not allow multiple selections, then the following command line can be used.

    List thelist = new List( );

In this command line, thelist is an instance of the list class.

Once the list has been created, the addItem method given below enables the user to add new entries.

    thelist.addItem("Coffee");

    thelist.addItem("Tea");

The user can also add an item in a specific location in the list.

   Example: thelist.addItem("Coffee", 0);

    In this example, the item **Coffee** is added to the first position(0) in **thelist.**

12.1.6 Scroll Bars

A scrollbar offers a user interface to scroll through a range of values. These values can then be connected with a variety of other uses. The scrollbar's maximum and minimum values can be initialized along with its minor changes such as the line increments, and major changes such as page increments.

The following statement can be used to create a horizontal scrollbar.

    Scrollbar colorscroll = new Scrollbar(Scrollbar. HORIZONTAL);

   Example: Scrollbar colorscroll = new Scrollbar(Scrollbar. VERTICAL, 0, 10, 2, 256);

    In this example, the line will create a vertical scrollbar with a starting position of 0, page size 10, minimum value 2, and maximum value of 256.

There are three different parts of a scrollbar that allow the user to select a value between the maximum and minimum. The increment or decrement value of the arrow can be set to a small unit. The default value is 1. The arrows increment or decrement with the line update can be set to a small unit and by default the value is 1. Clicking between the arrow and scroll box increases or decreases the value and the default value is 10. The box in the middle allows the user to click and drag to cross the scroll bar quickly from end to end.

12.1.7 Text Field

Text field is also a text container component of **java.awt** package. The text field component contains single line and limited text information.

The text field is declared as:

> TextField txtfield = new TextField(20);

The number of columns in the text field can be fixed by specifying the number in the constructor. In the above syntax, the number of columns is fixed to 20.

12.2 Layout Managers

Layouts are used to format components on the screen, which is platform-independent. This means that the programs can be executed in multiple platforms. Layout managers give programs a reliable and practical appearance, regardless of the platform, the screen size, or actions the user might take.

The java development kit offers five classes and these classes implement the layout manager interface. Figure 12.1 illustrates these five classes.

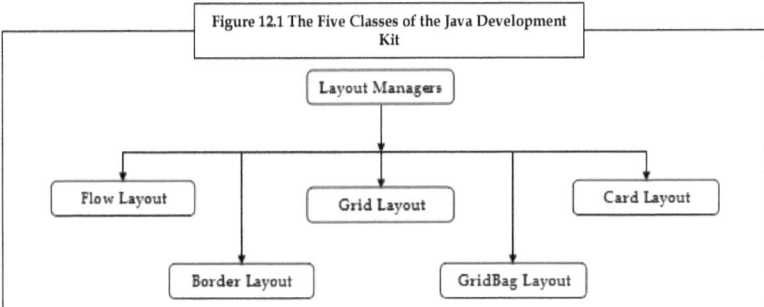

12.2.1 Flow Layout

Flow Layout is the default layout of the Panel class. When components are added to the screen, they flow from left to right based on the order added and the width of the applet. If there are many components that have to be placed in a window, then they wrap to a new row.

 Example: Program to illustrate the flow layout manager in Java.

```
import java.awt.*;
import javax. swing.*;
public class TestFlowLayout extends JApplet
{
JButton a1, a2, a3;
FlowLayout F;
public void init( )
{
 F= new FlowLayout(FlowLayout.LEFT);
 JPanel P = new JPanel( );
```

```
getContentPane( ).add(P);
P.setLayout (F);
a1= new JButton ("A");
a2= new JButton ("B");
a3= new JButton ("C");
P.add (a1);
P.add (a2);
P.add (a3);
}
}
```

HTML Coding:
```
<html>
<head></head>
<title></title>
<body>
<p>
<applet code="TestFlowLayout.class" width=250 height=250>
</applet>
</p>
</body>
</html>
```

Output:

In this example,

1. First, java.awt.* and javax. swing.* packages are imported.

2. Then, a class TestFlowLayout is created that extends the JApplet class present in the swing package.

3. In the class TestFlowLayout,

   (a) Variables **a1**, **a2** and **a3** are declared for the JButton class.

   (b) Then, a variable **F** is declared for the FlowLayout class.

   (c) The init( ) method is then called.

   (d) In the init( ) method,

      (i) The flow layout manager **F** is assigned the flow layout with the left alignment.

      (ii) A new instance **P** of the JPanel class is created.

      (iii) Then, the getContentPane( ) method is called to get a reference to the panel **P**.

      (iv) The setLayout( ) method is then called for the container **P** to specify that **F** layout is to be used to place components.

      (v) The variables **a1**, **a2** and **a3** are assigned the values of A, B and C.

      (vi) Then, these buttons **a1**, **a2** and **a3** are added to the panel **P** using add( ) method.

Task   Write a program to illustrate a flow layout manager in Java AWT. Drag the sides or corners of the displayed frame to demonstrate the working of the layout manager.

12.2.2 Border Layout

Border Layout is the default layout for Window, along with its children, Frame and Dialog Border Layout. Border layout offers five different areas to hold components. These five areas are named after the four borders of the screen, namely North, South, East and West. The remaining space is placed in the center area. Border Layout has one button in each area, before and after resizing.

   Example: Program to illustrate the border layout manager in Java.

```
import java.awt.*;
import javax.swing.*;
public class TestBorderLayout extends JApplet
{
JButton button1, button2, button3, button4, button5;
BorderLayout bl;
public void init( )
{
bl = new BorderLayout( );
JPanel pan = new JPanel( );
getContentPane( ).add(pan);
pan.setLayout(bl);
button1 = new JButton("North");
button2 = new JButton("South");
```

```
    button3 = new JButton("East");
    button4 = new JButton("West");
    button5 = new JButton("Center");
    pan.add("North", button1);
    pan.add("South", button2);
    pan.add("East", button3);
    pan.add("West", button4);
    pan.add("Center", button5);
  }
}
```

HTML Coding:

```
<html>
<head></head>
<title></title>
<body>
<p>
<applet code="TestBorderLayout.class" width=400 height=200>
</applet>
</p>
</body>
</html>
```

Output:

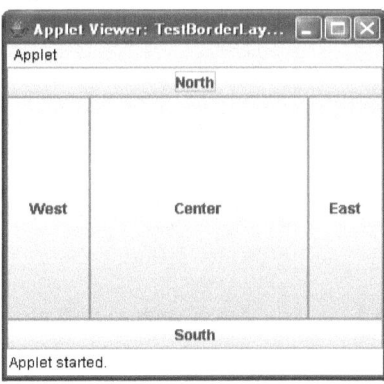

In this example,

1. First, the java.awt.* and javax.swing.* packages are imported.
2. Then, a class TestBorderLayout is created, which extends JApplet class.

3. In the class TestBorderLayout,
    (a) Five Jbuttons, namely, **button1**, **button2**, **button3**, **button4**, and **button5** are declared.
    (b) A variable **b1** is declared for BorderLayout.
    (c) Then, the init( ) method is called.
    (d) In the init( ) method,
        (i) A new BorderLayout is created and assigned to **b1**.
        (ii) Then, a new JPanel is created and assigned to the panel **pan**.
        (iii) The getContentPane( ) method is then called to get a reference to the panel **pan**.
        (iv) The setLayout( ) method is called to set the layout **b1** for the panel **pan**.
        (v) Then, a new JButton is created with the string North and is assigned to **button1**.
        (vi) A new JButton is created with the string South and is assigned to **button2**.
        (vii) Then, a new JButton is created with the string East and is assigned to **button3**.
        (viii) A new JButton is then created with the string West and is assigned to **button4**.
        (ix) Then, a new JButton is created with the string Center and is assigned to **button5**.
        (x) Finally, the JButtons **button1**, **button2**, **button3**, **button4** and **button5** are added to the panel **pan** using the add( ) method.

Caution When less than five components are placed in a container and **BorderLayout** is used, the empty component regions disappear. The remaining components then expand to fill the available space.

12.2.3 Grid Layout

Grid Layout helps in the arrangement of components in rows and columns. This process of arrangement of components starts at the first row and column, then moves across the row until it is full, and then continues on to the next row. Grid Layout Manager allows the user to reposition or resize objects after adding or removing components.

Example: Program to illustrate the grid layout manager in Java.

```
import java.awt.*;
import javax.swing.*;
public class TestGridLayout extends JApplet
{
JButton a1, a2, a3, a4, a5, a6;
GridLayout G;
public void init( )
```

```
{
  G=new GridLayout(3, 3);
  JPanel J = new JPanel ( );
  getContentPane( ).add(J);
  J.setLayout(G);
  a1=new JButton("A");
  a2=new JButton("B");
  a3=new JButton("C");
  a4=new JButton("D");
  a5=new JButton("E");
  a6=new JButton("F");
  J.add(a1);
  J.add(a2);
  J.add(a3);
  J.add(a4);
  J.add(a5);
  J.add(a6);
 }
}
```

HTML Coding:

```
<html>
<head></head>
<title></title>
<body>
<p>
<applet code="TestGridLayout.class" width=250 height=250>
</applet>
</p>
</body>
</html>
```

Output:

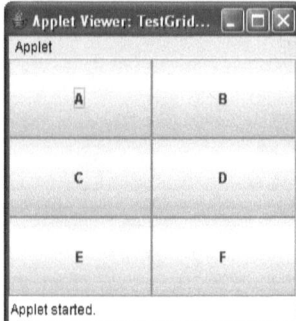

In this example,

1. First, java.awt.* and javax.swing.* packages are imported.

2. Then, a class TestGridLayout is created, which extends the JApplet class present in the swing package.

3. In the class TestGridLayout,

   (a) Six JButtons, namely, **a1**, **a2**, **a3**, **a4**, **a5** and **a6** are declared.

   (b) Then, a GridLayout **G** is declared.

   (c) The init( ) method is then called.

   (d) In the init( ) method,

      (i) A new GridLayout with dimensions 3, 3 is created and assigned to **G**.

      (ii) Then, a new JPanel instance is created and assigned to J.

      (iii) The getContentPane( ) method is then called to get a reference to the panel J.

      (iv) The setLayout( ) method is called to set the layout **G** for the panel J.

      (v) Then, the JButtons **a1**, **a2**, **a3**, **a4**, **a5** and **a6** are assigned values of A, B, C, D, E and F, respectively.

      (vi) Finally, these buttons are added to the panel J using the add( ) method.

12.2.4 GridBag Layout

GridBag Layout is complex when compared to the other layouts provided in the AWT. GridBag Layout allows the user to organize components in multiple rows and columns, stretch specific rows or columns when space is available, and anchor objects in different corners.

Example: Program to illustrate the GridBag layout manager in Java.

import java.awt.*;

import javax.swing.*;

```
public class TestGridBag extends JApplet
{
JPanel panelObj;
GridBagLayout gbObj;
GridBagConstraints gbCons;
public void init( )
{
  gbObj = new GridBagLayout( );
  gbCons = new GridBagConstraints( );
  panelObj = (JPanel)getContentPane( );
  panelObj.setLayout(gbObj);
  JButton button1 = new JButton("Button 1");
  JButton button2 = new JButton("Button 2");
  JButton button3 = new JButton("Button 3");
  JButton button4 = new JButton("Button 4");
  gbCons.fill = GridBagConstraints.BOTH;
  gbCons.anchor = GridBagConstraints.CENTER;
  gbCons.gridwidth = 1;
  gbCons.weightx = 1.0;
  gbObj.setConstraints(button1, gbCons);
  panelObj.add(button1);
  gbCons.gridwidth = GridBagConstraints.REMAINDER;
  gbObj.setConstraints(button2, gbCons);
  panelObj.add(button2);
  gbCons.gridwidth = GridBagConstraints.REMAINDER;
  gbObj.setConstraints(button3, gbCons);
  panelObj.add(button3);
  gbCons.fill = GridBagConstraints.BOTH;
  gbCons.anchor = GridBagConstraints.CENTER;
  gbCons.gridwidth = 1;
  gbCons.weightx = 1.0;
  gbObj.setConstraints(button4, gbCons);
  panelObj.add(button4);
}
}
```

HTML Coding:

```
<html>
<head></head>
<title></title>
<body bgcolor=pink>
<p>
<h2>
<font size=30>Applet
</font>
<applet code="TestGridBag.class" width=250 height=250>
</applet>
</p>
</body>
</html>
```

Output:

In this example,

1. First, java.awt.* and javax.swing.* packages are imported.

2. Then, a class **TestGridBag** is created, which extends JApplet class.

3. In the class **TestGridBag**,

    (a) A JPanel object **panelObj** is declared.

    (b) Then, GridBagLayout object **gbObj** is declared.

    (c) The GridBagConstraints are then declared by using **gbCons**.

    (d) The init( ) method is then called.

    (e) In the init( ) method,

        (i) A new instance of the GridBag layout manager is created and
             assigned to **gbObj**.

(ii) A new instance of the GridBagConstraints layout manager is created and assigned to **gbCons**.

(iii) Then, panelObj = (JPanel)getContentPane( ); statement is used to get the contentpane value and assign it to **panelObj**.

(iv) The setLayout( ) method is called to set the layout **gbObj** for the panel **panelObj**.

(v) Then, new instances of JButtons are created with values Button 1, Button 2, Button 3 and Button 4. These values are assigned to **button1, button2, button3,** and **button4** respectively.

(vi) The fill attribute is used when a component is smaller than its display area for ascertaining whether the component needs to be stretched within its display area. The GridBagConstraints.BOTH fills the display area completely.

(vii) The anchor attribute is used when a component is smaller than its display area. It is used for determining where the component has to be placed in the display area. GridBagConstraints.CENTER is the default value.

(viii) The gridwidth and weightx of **gbCons** are assigned values of 1 and 1.0, respectively. The gridwidth attribute specifies the number of columns to be used as the display area of the component. The attribute weightx specifies whether the components stretch horizontally to fill the display area of the applet.

(ix) The setConstraints( ) method of the GridBagLayout class is used for associating the constraints with the component.

(x) Then, the **button1** is added to the **panelObj** using the add( ) method.

(xi) The remainder value of GridBagConstraints is assigned to the gridwidth of **gbCons**.

(xii) Then, setConstraints( ) method is called to set the constraints **button2** and **gbCons** of the object **gbObj**.

(xiii) The **button2** is then added to **panelObj** using the add( ) method.

(xiv) The remainder value of GridBagConstraints is assigned to the gridwidth of **gbCons**.

(xv) Then, setConstraints( ) method is called to set the constraints **button3** and **gbCons**, of the object **gbObj**.

(xvi) The **button3** is then added to **panelObj** using the add( ) method.

(xvii) The GridBagConstraints.BOTH fills the display area completely and assigns the result to gbCons.fill.

(xviii) Then, GridBagConstraints.CENTER anchor attribute is used as the default value of **gbCons**.

(xix) The gridwidth and weightx of **gbCons** are assigned the values of 1 and 1.0, respectively.

(xx) Then, setConstraints( ) method is called to set the constraints **button4** and **gbCons** of the object **gbObj**.

(xxi) Finally, the **button4** is then added to **panelObj** using the add( )
method.

12.2.5 Card Layout

A card layout is used to manage several components. All the components are provided the same size.
The card layout is used to manage a group of panels and these panels have their own components.

Example: Program to illustrate the card layout manager in Java.

```
import java.awt.*;
import javax.swing.*;
public class TestCardLayout extends JApplet
{
JButton a1, a2, a3;
CardLayout C;
JPanel p;
public void init( )
{
  p=new JPanel( ); C=new
  CardLayout( );
  p.setLayout(C);
  getContentPane( ).add (p);
  a1=new JButton("a");
  a2=new JButton("b");
  a3=new JButton("c");
  p.add("a",a1);
  p.add("b",a2);
  p.add("c",a3);
 }
}
```

HTML Coding:

```
<html>
<head></head>
<title></title>
<body>
<p>
<applet code="TestCardLayout.class" width=400 height=250>
</applet>
```

```
</p>
</body>
</html>
```

Output:

In this example,

1. First, java.awt.* and javax.swing.* packages are imported.

2. Then, a class TestCardLayout is created that extends the JApplet class.

3. In the class TestCardLayout,

   (a) Three JButtons, namely, **a1**, **a2** and **a3** are declared.

   (b) Then, an object **C** is declared for the CardLayout class.

   (c) A variable **p** is used to declare a JPanel object.

   (d) Then, the init( ) method is called.

   (e) In the init( ) method,

       (i) A new instance of the JPanel class is created and assigned to **p**.

       (ii) A new instance of the CardLayout class is created and assigned to **C**.

       (iii) Then, the setLayout( ) method is called to set the layout **C** for panel **p**.

       (iv) The getContentPane( ) method is then called to get a reference to the panel **p**.

       (v) Then, new instances of JButtons are created with values a, b and c, and are assigned to **a1**, **a2** and **a3**, respectively.

       (vi) Finally, these buttons and their values are then added to the panel **p** using the add( ) method.

Lab Exercise   Write a program to illustrate a grid layout manager in Java AWT. The grid layout must contain five grids with different names.

NAGESH JAITAK

12.3 Summary

*   AWT is a class library offered by Java programming language, which provides a common set of tools to design GUIs that is platform-independent.

*   The **java.awt.*** package is used to develop the user interface objects such as check boxes and radio buttons.

*   Layouts are used to format components on the screen.

*   A good appearance, regardless of the platform, screen size, or actions the user might take is given by the layout managers.

*   The different types of layout managers are flow layout, border layout, grid layout, gridbag layout and card layout.

12.4 Key words

Class Library: A large body of code that applications can call at runtime.

Container: A component that contains other components within it.

Panel: Simplest container class that provides space in an application to attach other components, which may also include other panels.

Super Class: A class that provides a method or methods to a Java subclass.

12.5 Self Assessment

1. State whether the following statements are true or false:

    (a) A label is a simple component of java AWT.

    (b) Card Layout is the default layout of the Panel class.

    (c) GridBag Layout allows the user to manage a group of Panels and these panels have their own components.

2. Fill in the blanks:

    (a) Text field is a text container component of _____package.

    (b) The _____layout helps in the arrangement of components in rows and columns.

    (c) A _____layout is used to manage several components.

3. Select a suitable choice in every question.

    (a) Which of the following classes permits a user to create a scrolling list of values that can be selected alone or together?
        (i)   Scrollbar class
        (ii) Choice class
        (iii) List class
        (iv) Checkbox class

    (b) Which of the following is the default layout for Window?
        (i)   Border layout
        (ii) Flow layout
        (iii) Grid layout
        (iv) Card layout

12.6 Review Questions

1. "AWT has a collection of basic user interface components". Justify.
2. "The Java development kit offers five classes and these classes implement the layout manager

interface". Discuss.

3. "When components are added to the screen, they flow from left to right in flow layout". Illustrate this using a program.

4. "GridBag Layout allows the user to organize components in multiple rows and columns, stretch specific rows or columns when space is available, and anchor objects in different corners". Illustrate this using a program.

5. "The Card layout is used to manage a group of panels and these panels have their own components". Illustrate this using a program.

Answers: Self Assessment

1. (a) True   (b) False   (c) False

2. (a) java.awt   (b) Grid   (c) Card

3. (a) List class   (b) Border layout

12.7 Further Readings

Books

Balagurusamy E. Programming with Java_A Primer 3e. New Delhi

Schildt. H. Java 2 The Complete Reference, 5th ed. New York: McGraw-Hill/Osborne.

Online link

http://www.roseindia.net/java/example/java/awt/AwtComponents.shtml

http://www.jhlabs.com/java/layout/index.html

Unit 13: Swings

Objectives

After studying this unit, you will be able to:

- Explain swing components

- Discuss various GUI components

- Compare the model of buttons, checkboxes and radio buttons

- Use combo box and lists in Java

- Analyze the concept of menus in Java

Introduction

Swing is an important component of Java language, which is provided in the **javax.swing** package. The concept of swings came into existence for the purpose of providing a more advanced collection of GUI (Graphical User Interface) components to Java as compared to the AWT.

Did you know? The Swing library was introduced by Sun Microsystems to create a GUI, which is elegant, object-oriented and easy to work with. Swing library is contained in javax.swing and java.awt packages.

A Swing is a graphical user interface library of Java. It provides the multiple platform independent API interfaces. These interfaces are used for interacting between the users and GUI components. Swing is a part of Java Foundation Class (JFC) and it includes the graphical widgets like checkboxes, radio buttons, menus, and so on. The Java Swing can handle all the AWT flexible components. Swing and its components are commonly used in Java because windows' appearance can be changed easily using an important feature of Swings, that is, pluggable look and feel. The **pluggable look and feel** of Java Swing is a mechanism that permits the modification in the look and feel of the GUI at runtime.

13.1 Features of Swings

The Java Abstract Window Toolkit (AWT) provides a platform-specific code. However, Swing is written in Java and hence is platform independent. Unlike the AWT, Swing has more sophisticated interface capabilities. Swing offers features like tabbed panes. It also has the ability to change images on buttons.

Caution Swing and AWT are incompatible. These cannot be mixed. Only one can be used at a time.

Few advantages of Swings that make them user friendly are their extensibility, customizability and configurability. Swing components are lightweight, and hence provide a better user interface.

The names of the components of Swings start with the letter J.

 Example: JButton, JLabel, and JSlider.

The **JComponent** (a Swing component), which is derived directly from container, is the foundation class for most of the user interface components of a Swing. The **container class** is a class in Java which is used to add components in the container and laying them.

Did you know? In Java, there are 250 new classes and 75 interfaces in Swing.

 Javax.swing.SwingWorker class is the generic solution to the issues of updating the *Notes* **GUI from worker threads and giving users the ability to control the background tasks.**

1. This class provides a means of returning intermediate results from the background task.

2. This class provides a method for updating the interface with the intermediate results.

3. This class solves memory inconsistency errors.

Figure 13.1 depicts the swing class hierarchy, which comprises a group of swing components.

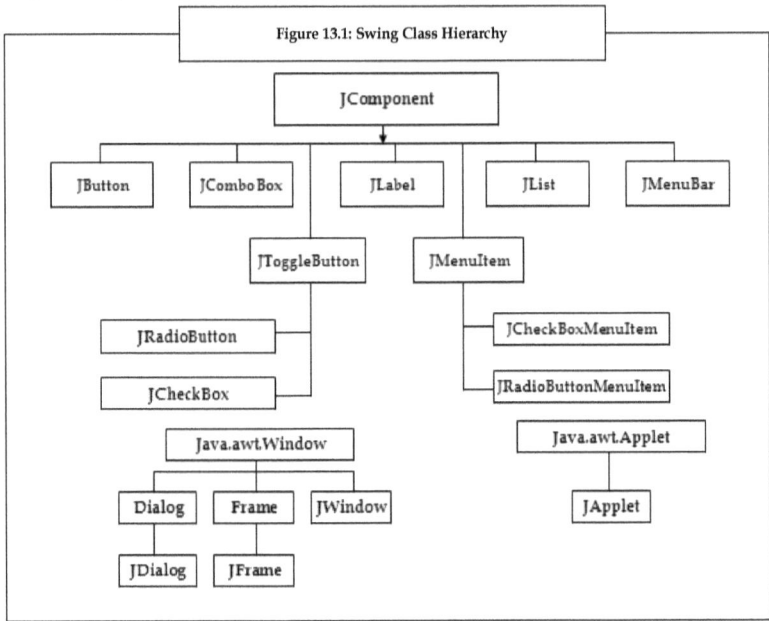

In the figure 13.1, **JComponent** class is derived from the container class of AWT. Therefore, the swing class properties are inherited from the component class and container class. The **component class** is a Java class, which offers the layout hints and assists the painting and event hints. The **JComponent** class consists of JButton, JComboBox, JLabel, JList, JMenuBar, JToggleButton and the JMenuItem component. These components are further classified into various other components.

Caution

Fields must be created only if necessary. If a particular element is created, wired into the GUI, and then no longer referenced anywhere else in the class, then it should probably not exist as a field, but rather as a simple local variable. This style will allow the more important GUI elements (fields) to stand out from the less important ones (local variables).

Swings have 18 public packages.

Example: Program to illustrate the concept of swing in Java.

```
import javax.swing.*;

import java.awt.*;

public class Sample extends JApplet
{
    public void init( )
```

```
    {
        Container contentPane = getContentPane( );
        JLabel j= new JLabel ("Sample");
        contentPane.add(j);
    }
}
```

HTML Coding:

```
<html>
<head></head>
<title></title>
<body >
<p>
<applet code="Sample.class" width=250 height=250>
</applet>
</p>
</body>
</html>
```

Output:

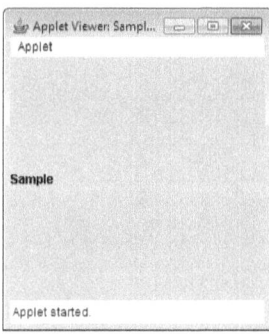

In this example,

1.   First, a class **Sample** is created from the **JApplet** class, by using the **extends** keyword, and is declared public by using the **public** keyword.

2.   In the class **Sample**,

      (a) The **init( )** method is called, which is declared **public** and **void.**

      (b) In this **init( )** method,

            (i) The **Container contentPane = getContentPane( );** statement is used for adding the component to the container's content pane.

(ii) Then, a new JLabel **Sample** is created, and is assigned to JLabel **j**.

(iii) Finally, the statement **contentPane.add(j);** is used for adding the JLabel j to the content pane.

Note: Finally, the code is executed in the command prompt, and for acquiring the output, an HTML code is written. The result is shown as an applet window with a label Sample on it.

13.2 Swing Components

As the concept of Swings is important in Java, so are their components, which are considered to be the basic building blocks of an application using Swings. These components are contained in the Swing packages. Swing packages are similar to the AWT packages and are used to provide classes for creating GUI applications. The **javax.swing** package contains all the Swing components. The only difference between the AWT and Swing packages is that the Swing packages are completely written in Java, whereas the AWT packages are not completely written in Java. As a result, the GUI programs, which are written using classes from the Swing package, have a similar appearance when executed on different platforms.

Table 13.1 describes the different Swing components.

Table 13.1: Swing Components

Swing Component	Description
JComponent	This component is the root class for all Swing components, but not for the top-level containers.
JButton	This component is known as a push button, which resembles the Button class in the AWT package.
JCheckBox	This component is used to create a code, where the user can select or deselect an item. This component looks like the Checkbox class in the AWT package.
JFileChooser	This component permits the user to select a file, and it resembles the FileChooser class in the AWT package.
JTextField	This component permits the single-line text editing, and corresponds to TextField class in the AWT package.
JFrame	This component extends and matches the Frame class in the AWT package. Though, the two are marginally mismatched in terms of adding components to this container.
JPanel	This component extends JComponent and resembles the Panel class in the AWT package.
JApplet	This component extends and matches the Applet class in the AWT package. But, it is not completely compatible with the Applet class in terms of adding components to this container.
JOptionPane	This swing component extends JComponent. It is used to display a pop-up dialog box.

Cont..

JDialog	This component extends and resembles the dialog class in the AWT package. It is used to inform the user of something or prompt the user for an input.
JColorChooser	This component extends JComponent and allows the user to select a color.

Let us next discuss some of the main swing components given in the table 13.1.

Frames, Panels and Applets

Frames, panels, and applets are the components of Swing class.

1. *JFrame:* Frames use the content pane (class) to add all the components. A user can change the properties of **JFrame** (a Swing component), such as, changing the layout manager, background color, and so on. These changes have an effect on the content pane. Frames can be closed by just clicking on the close button. This would mean that only the window is closed, not the program. A user can use the **setDefaultCloseOperation (EXIT_ON_CLOSE)** method, which allows the user to close the **JFrame**.

2. *JPanel:* The JPanel component is an intermediate container, and a Swing component that is used to group the lightweight swing components. The **JPanel** works in the same way as the content pane does. A **FlowLayout** is the default layout for a **JPanel**.

3. *JApplet:* The JApplet is a Swing component and contains a content pane. The components can be added in this content pane. The layout, background color and several other properties of **JApplet** can be changed, and these changes also apply to the content pane. The default layout manager for **JApplet** is the **BorderLayout**.

Labels

As frame, panels, and applets are Swing components, similarly labels are also Swing components. The idea behind using a **JLabel** is to display an area for a short text string or an image, or both. Labels do not respond to input events; hence, **JLabel** does not receive the keyboard focus. The label can be aligned by setting the label's contents using the vertical and horizontal alignment.

 Example: Labels are vertically centered in their display area, by default. The labels, which contain only text, are aligned at the leading edge, by default; and the labels, which contain only images, are horizontally centered. But generally, the text is on the trailing edge of the image, where the text and image are vertically aligned.

Some pixels must be given between the label text and the image, and for specifying these pixels, the setIconTextGap method is used. The default pixel is 4.

 Example: Program to illustrate the concept of labels in Java.

```
import java.awt.*;
import java.awt.event.*;
import javax.swing.*;
public class JLabelDemo extends JApplet
{
```

```
private Container Panel;
private LayoutManager Layout;
private JLabel Label1;
private JLabel Label2;
private JLabel Label3;
private JLabel Label4;
private JLabel Label5;
private JLabel Label6;
private JLabel Label7;
public JLabelDemo( )
{
Layout = new GridLayout (7, 1);
Label1 = new JLabel ("A Simple Label");
Label2 = new JLabel ("A Label with LEFT alignment", JLabel.LEFT);
Label3 = new JLabel ("A Label with CENTER alignment", JLabel.CENTER);
Label4 = new JLabel ("A Label with RIGHT alignment", JLabel.RIGHT);
Label5 = new JLabel ("A Label with LEADING alignment",
JLabel.LEADING);
Label6 = new JLabel ("A Label with TRAILING alignment", JLabel.TRAILING);
Label7 = new JLabel( );
Panel = getContentPane( );
Panel.setLayout (Layout);
Panel.add (Label1);
Panel.add (Label2);
Panel.add (Label3);
Panel.add (Label4);
Panel.add (Label5);
Panel.add (Label6);
Panel.add (Label7);
Panel.setBackground (Color.gray);
Label7.setHorizontalAlignment(JLabel.CENTER);
Label7.setForeground(Color.blue);
Label7.setText("Text added with setText");
}
}
```

HTML Coding:

<html>

<head></head>

<title></title>

<body>

<p>

<applet code="JLabelDemo.class" width=250 height=250>

</applet>

</p>

</body>

</html>

Output:

In this example,

1. First, a class **JLabelDemo** is created from the class **JApplet**, by using the **extends** keyword.

2. In the class **JLabelDemo**,

    a. First, a container panel, a layout manager Layout, and seven labels from **Label1** to **Label7** are declared as JLabel.

    b. Then, the JLabelDemo( ) constructor is called. In this constructor,

        (i) The Layout = new GridLayout (7, 1); statement, GridLayout is created by using the new keyword, and is linked to **Layout**.

        (ii) Different JLabels are created by using the new keyword, and are linked from **Label 1** to **Label 7**, respectively.

(iii) The Panel = getContentPane( ); statement is used to get the content pane and link it to **Panel**.

(iv) Then, the layout (here, GridLayout) of the panel is set using the setLayout( ) method in the Panel.setLayout (Layout); statement.

(v) All the labels that were created in the program are then added to **Panel**, by using the add( ) method.

(vi) Then, the background color of **Panel** is set to gray color, by using the setBackground( ) method, in the Panel.setBackground (Color.gray); statement.

(vii) Finally, the horizontal alignment, foreground, and text of **Label7** are set, by using the setHorizontalAlignment( ), setForeground( ), and setText( ) methods, respectively.

Note: Finally, the code is executed in the command prompt, and for acquiring the output, an HTML code is written. The result is shown as an applet window and labels with different names on it.

Text Components

Text components are also Swing components. The Swing text components are used to display text. The text components permit the editing of the text, to the user. Java Swings contain six text components, supporting classes and interfaces, which can be used to write complex codes. All the swing text components have different uses and are inherited from the super class **JTextComponent**. **JTextComponent** helps the user in making the foundation for text manipulation highly-configurable and powerful.

Figure 13.2 depicts the **JTextComponent** hierarchy, which contains different text components, like
JTextField, JtextArea and JEditorPane.

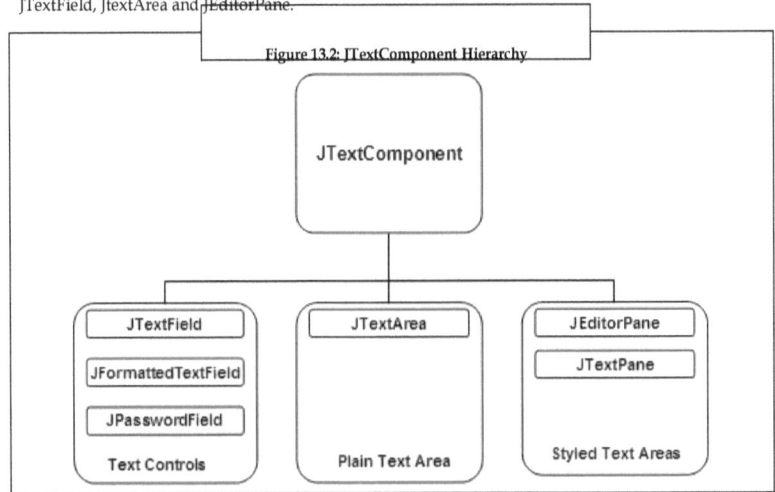

Figure 13.2: JTextComponent Hierarchy

As given in the figure 13.2, **JTextComponents** can be classified into three sections, namely, Text Controls, Plain Text Areas, and Styled Text Areas.

1. **Text controls:** The Text control section is one of the sections of the **JTextComponent**. The text controls contain three components, namely, JTextField, JFormattedTextField and JPasswordField.

    (a) *JTextField*: This is a swing text component. This component is known as a lightweight component and allows the user to edit a single line of text.

    (b) *JFormattedTextField*: This component allows the user to format arbitrary values. It is also used to retrieve objects after the text has been edited.

    (c) *JPasswordField*: This component is known as a lightweight component in Swings. This component allows the user to edit a single line of text, where the view specifies the typed text not the original text.

2. **Plain Text Areas:** The plain text area is another section of the **JTextComponent**. It includes the component named as **JTextArea**.

    (a) *JtextArea*: This is a swing text component. The **JtextArea** component is used to display plain text and is a multi-line text area. It is known as a lightweight component for working with text. This text component does not handle scrolling, and for this reason **JScrollPane** component is used.

3. **Styled Text Area:** The styled text area is the third section of the **JTextComponent,** which contains the component **JTextPane**.

    (a) *JTextPane Component*: The **JTextPane** component is an advanced component that is used for working with text. This component is used to write codes, which contain complex formatting operations. **JTextPane** can also exhibit HTML documents.

All the above components have their own importance in handling the Swing text.

 Example: Program to illustrate JTextArea component in Java.

```
import java.awt.*;
import java.awt.event.*;
import javax.swing.*;
public class TextArea extends JApplet implements ActionListener
{
    JTextField Sample;
    JTextArea e1;
    JTextArea e2;
    Container Panel;
    LayoutManager Layout;
    public TextArea( )
    {
        Sample = new JTextField ("Sample ", 20);
        e1 = new JTextArea (5, 20);
        e2 = new JTextArea (5, 20);
```

```
Layout = new FlowLayout( );
Panel = getContentPane( );
Sample.addActionListener (this);
e1.setEditable (false);
e1.setBackground (Color.white);
e1.setLineWrap (true);
Panel.setLayout (Layout);
Panel.add (Sample);
Panel.add (e1); Panel.add (e2);
Panel.setBackground(Color.pink);
}
public void actionPerformed(ActionEvent e)
{
  String Reply;
  Reply = "The text which was entered into the JTextField was \"" +
        Sample.getText( ) + "\", and this is the echo with text wrap.";
  e1.setText (Reply);
  Reply = "The text which was entered\ninto the JTextField was\n\""
        + Sample.getText( )
      + "\", \and this is the echo with returns";
  e2.setText (Reply);
  }
}
```

HTML Coding:

```
<html>
<head></head>
<title></title>
<body>
<p>
<applet code="TextArea.class" width=225 height=200>
</applet>
</p>
</body>
</html>
```

Output:

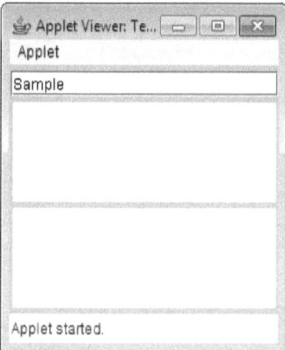

In this example,

1. First, a class **TextArea** is created, which **extends** the class **JApplet** and **implements** the class **ActionListener.**

2. In this class,

    (a) A JTextField **Sample**, two JTextAreas **e1** and **e2**, Container **Panel**, and LayoutManager **Layout**, are declared.

    (b) Then, the TextArea( ) constructor is called. In this constructor,

        (i) A new **JTextField** is created, and is linked to the **Sample.**

        (ii) Two new **JTextAreas** are created and linked to **e1** and **e2**, respectively.

        (iii) A new **FlowLayout** is created, and is linked to the **Layout.**

        (iv) Then, **getContentPane( )** method is called to get the content pane, and is linked to the **Panel.**

        (v) Afterwards, the **addActionListener (this)** method is added to the **Sample.**

        (vi) The editable field, background, and the line wrap of the text area **e1** are set, by using the respective methods.

        (vii) Then, **SetLayout** method is called to set the layout of the **Panel.**

        (viii) The **Sample, e1**, and **e2** are then added to **Panel**, by using the **add( )** method.

        (ix) Then, the background of the **Panel** is set to pink color, by using the **setBackground( )** method.

    (c) Then, the **ActionEvent e** of the **actionPerformed( )** method is called.

    (d) In this method,

        (i) First, a String Reply is declared.

        (ii) Then, the string Reply is assigned a string value.

        (iii) The text assigned to string Reply of e1 is then set, by using the setText( ) method.

(iv) Then, again the string Reply is assigned a new string value.

(v) The text assigned to string Reply of e2 is then set, by using the setText( ) method.

Note: Finally, the code is executed in the command prompt. For acquiring the output, an HTML code is written. The result is shown as an applet window, and the text area and text field on it.

Buttons, Checkboxes and Radio Buttons

Apart from the Swing components discussed above, there are few more Swing components, which are the buttons, checkboxes and radio buttons. The buttons are the classes derived from the **AbstractButton** class. The **AbstractButton** class is a class, which contains different methods that control the behavior of buttons, checkboxes and radio buttons.

1. **Buttons:** Buttons are simple buttons, on which a user can click and perform the desired operations. **JButtons** are She swing components which extend the **JComponent**. The button class in Swings is similar to the button class found in **java.awt.Button** package. The buttons can be arranged and organized by **Actions**. Swing buttons are used to display both the text and an image. The letter that is underlined in the text of each button is the keyboard alternative for each button. The appearance of the button is auto-generated with the button's disabled appearance. The user can provide an image as a substitute for the normal image.

Notes   A user must implement an action listener while creating buttons.

The following example will clarify the concept of buttons in Java.

Example: Program to illustrate button component in Java.

```
import java.awt.*;
import javax.swing.*;
import java.awt.event.*;
public class ButtonSample extends JApplet implements ActionListener
{
    JTextField g;
    public void init( )
    {
    Container contentPane = getContentPane( );
    contentPane.setLayout(new FlowLayout( ));
    JButton b = new JButton("Rose");
    b.setActionCommand("Rose") ;
    b.addActionListener(this);
    contentPane.add(b);
```

```
b = new JButton("Lily");
b.setActionCommand("Lily") ;
b.addActionListener(this);
contentPane.add(b);
b = new JButton("Lotus");
b.setActionCommand("Lotus");
b.addActionListener(this);
contentPane.add(b);
g = new JTextField(20);
 contentPane.add(g);
}
 public void actionPerformed(ActionEvent e)
 {
    g.setText(e.getActionCommand( ));
 }
}
```

HTML Coding:

```
<html>
<head></head>
<title></title>
<body>
<p>
<applet code="ButtonSample.class" width=225 height=200>
</applet>
</p>
</body>
</html>
```

Output:

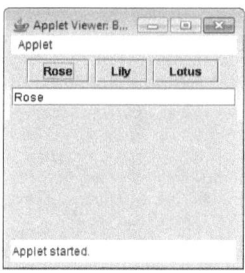

In this example,

1. First, a class **ButtonSample** is created, which extends the class JApplet, and implements the class ActionListener.

2. In this class,

    (a) A JTextField object g is declared, and the init( ) method is called.

    (b) In this method,

        (i)   The getContentPane( ) method is called to get the content pane; this content pane is linked to the container **contentPane**.

        (ii) A new FlowLayout is created while setting the layout of the content pane with the help of the setLayout( ) method.

        (iii) Then, a new JButton is created with the string **Rose**, and is linked to the JButton **b.**

        (iv) The setActionCommand( ) method is called to set the content of b.

        (v)   Then, the addActionListener( ) method is called to on the JButton **b.**

        (vi) The string in b is then added to the content pane, by using the add( ) method.

        (vii) Another new JButton with string **Lily** is created, and is linked to **b.**

        (viii) Repeat steps (iv) to (vi) for the new value stored in **b.**

        (ix) Another new JButton with string **Lotus** is created, and is linked to **b.**

        (x)   Repeat steps (iv) to (vi) for the new value stored in **b.**

        (xi) A new JTextField of value **20** is created, and is linked to **g.**

        (xii) Then, actionPerformed( ) method for ActionEvent **e**, is called, and g.setText(e.getActionCommand( )); method is also called within this method.

    Note: Finally, the code is executed in the command prompt. For acquiring the output, an HTML code is written. The result is shown as an applet window with the text area and text field in the applet.

Task   Write a simple Java program to illustrate the concept of JButtons, and JTextFields.

---

2. **CheckBox:** Just like Buttons, CheckBoxes are also Swing components. A Checkbox is defined as an item which a user can select or deselect with a single click. The checkbox item gives the user, the privilege to select a number of checkboxes from a group or the user may select all of them. The **JCheckBox** is a component of swing. The **JCheckBox** class assists the checkbox buttons. **JCheckBox** is a control that allows the user to select more than one attribute at a time by checking, that is ticking selections in a list. This is beneficial when multiple choices are involved.

All the CheckBoxes are created in such a way, that they are either selected or deselected. The checkboxes can also be included in menus. For this, the **JCheckBoxMenuItem** class (discussed later) is used. Swing checkboxes have common button characteristics, since JCheckBox and JCheckBoxMenuItem are inherited from **Abstract Button** class. Any number of checkboxes can be selected in a group, or all can be selected at once. On every click on the check box, it creates one item event and one action event.

Example: Program to illustrate the checkbox component in Java.

```
import java.awt.*;
import javax.swing.*;
import java.awt.event.*;
public class CheckBoxDemo extends JApplet implements ActionListener
{
  JTextField g;
  public void init( )
  {
    Container contentPane = getContentPane( );
    contentPane.setLayout(new  FlowLayout( ));
    JCheckBox c = new JCheckBox("Movies");
    c .setActionCommand("Movies") ;
    c.addActionListener(this);
    contentPane.add(c);
    c = new JCheckBox("Sports");
    c.setActionCommand("Sports") ;
    c.addActionListener(this);
    contentPane.add(c);
    c = new JCheckBox("Politics");
    c.setActionCommand("Politics") ;
    c.addActionListener(this);
    contentPane.add(c);
    g = new JTextField(20);
    contentPane.add(g);
  }
  public void actionPerformed(ActionEvent e)
  {
    g.setText(e.getActionCommand( ));
  }
}
```

HTML Coding:

```
<html>
<head></head>
<title></title>
<body>
```

```
<p>
<applet code="CheckBoxDemo.class" width=225 height=200>
</applet>
</p>
</body>
</html>
```

Output:

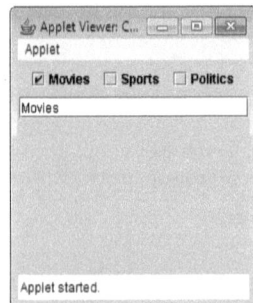

In this example,

1. First, a class **CheckBoxDemo** is created, which extends the class JApplet, and implements the class ActionListener.

2. In this class,

    (a) A JTextField object **g** is declared, and the init( ) method is called.

    (b) In this method,

        (i)   The getContentPane( ) method is called to get the content pane; this content pane is linked to the container **contentPane**.

        (ii) A new FlowLayout is created while setting the layout of the content pane with the help of setLayout( ) method.

        (iii) Then, a new JCheckBox is created with the string **Movies**, and is linked to the JCheckBox **c**.

        (iv) The setActionCommand( ) method is called to set the content of **c**.

        (v)   Then, the addActionListener( ) method is called on the JCheckBox **c**.

        (vi) The string in **c** is then added to the content pane, by using the add( ) method.

        (vii) Another new JCheckBox with string **Sports** is created, and is linked to **c**.

        (viii) Repeat steps (iv) to (vi) for the new value stored in **c**.

        (ix) Another new JCheckBox with string **Sports** is created, and is linked to **c**.

        (x)   Repeat steps (iv) to (vi) for the new value stored in **c**.

        (xi) A new JTextField of value **20** is created, and is linked to **g**.

(xii) Then, actionPerformed( ) method for ActionEvent **e**, is called, and g.setText(e.getActionCommand( )); method is also called within this method.

Note: Finally, the code is executed in the command prompt. For acquiring the output, an HTML code is written. The result is shown as an applet window with the text area and text field in the applet.

3. *Radio Buttons:* A Radio button is also a Swing component, which is defined as an item that a user can select or deselect with a single click. From a group of radio buttons, only one can be selected per click unlike check boxes. Prior to creation of any radio buttons, the user must create an instance of that button, the **ButtonGroup** object. Then, the user must add the radio buttons in the **ButtonGroup** object.

As a Swing component, a radio button is written as **JRadioButton**. A JRadioButton object can be created using many constructors. Some of these constructors are:

(a) JRadioButton( )

(b) JRadioButton (Icon icon)

(c) JRadioButton (Icon icon, Boolean selected)

(d) JRadioButton (String text)

(e) JRadioButton(String text, Boolean selected)

(f) JRadioButton (String text, Icon icon)

(g) JRadioButton (String text, Icon icon, Boolean selected)

ComboBoxes and Lists

1. *ComboBoxes:* The combobox and list box are Swing components which provide the user an option to select a number of item from a large list.

A Swing component, which brings a button or editable field and a drop-down list together, from which, the user can select any value, is referred to as a **ComboBox**. The **ComboBox** can be made editable also. For doing this, an editable field is added into the **ComboBox**, in which, the user can type a value.

A **ComboBox** contains three components, namely, a text field, a button, and a list. These components together perform some functions. They help in:

(a) Displaying the selected text field on the screen.

(b) Exhibiting the display of the list box that is controlled by the button that is present on the right side of the text field.

(c) Editing the selected text field.

(d) Displaying icons along with or in the place of the text.

 Example: Program to illustrate the combobox component in Java.

```
import java.awt.*;

import java.awt.event.*;

import javax.swing.*;

public class ComboBoxExample extends JApplet implements ItemListener

{
```

```
JLabel j;
ImageIcon ROSE, LILY, LOTUS, CARNATION, MARIGOLD;
public void init( )
{
  Container contentPane = getContentPane( );
  contentPane.setLayout(new FlowLayout( ));
  JComboBox c= new JComboBox( );
  c.addItem("ROSE");
  c.addItem("LILY");
  c.addItem("LOTUS");
  c.addItem("CARNATION");
  c.addItem("MARIGOLD");
  contentPane.add(c);
  j=new JLabel(new ImageIcon("ROSE.gif"));
  contentPane.add(j);
}
public void itemStateChanged(ItemEvent e)
{
  String s=(String)e.getItem( );
  j.setIcon(new ImageIcon(s+".gif"));
}
}
```

HTML Coding:

```
<html>
<head></head>
<title></title>
<body>
<p>
<applet code="ComboBoxExample.class" width=225 height=200>
</applet>
</p>
</body>
</html>
```

Output:

In this example,

1. First, a class **ComboBoxExample** is created, which extends the class JApplet, and implements the class ItemListener.

2. In this class,

    (a) A JLabel object **j** is declared,

    (b) The ImageIcon objects ROSE, LILY, LOTUS, CARNATION, MARIGOLD are declared and the **init( )** method is called.

    (c) In this method,

        (i) The getContentPane( ) method is called to get the content pane; this content pane is linked to the container **contentPane**.

        (ii) A new FlowLayout is created while setting the layout of the content pane with the help of the setLayout( ) method.Then, a new JComboBox is created, and is linked to the JComboBox c.

        (iii) The string **ROSE** is added to **c** by using the addItem( ) method.

        (iv) The string **LILY** is added to **c** by using the addItem( ) method.

        (v) The string **LOTUS** is added to **c** by using the addItem( ) method.

        (vi) The string **CARNATION** is added to **c** by using the addItem( ) method.

        (vii) The string **MARIGOLD** is added to **c** by using the addItem( ) method.

        (viii) The string in **c** is then added to the content pane, by using the add( ) method.

        (ix) A new JLabel is created with the ImageIcon **("ROSE.gif")**, and is linked to **j**.

        (x) The string in **j** is then added to the content pane, by using the add( ) method.

        (xi) The itemStateChanged( ) method is called where itemevent **e** is defined.

NAGESH JAITAK

Note: Finally, the code is executed in the command prompt. For acquiring the output, an HTML code is written. The result is shown as an applet window with the text area and text field in the applet.

2. **Lists:** A list is a simple presentation of choices, which are large in number. As a Swing component, list is written as **JList**. The **JList** component is defined as a component, which provides a set of items that are scrollable, from which, one or more items may be selected. For creating a **JList** component, the user must associate the list component with the scroll pane, since the **JList** does not support scrolling directly. The **Jlist** actions are handled by **ListSelectionListener** class.

The **JList** component can be created by using the below-given constructors:

(a) *public JList( ):* This constructor creates a **JList** with an empty model.

(b) *public JList (ListModel dataModel):* This constructor creates a **JList** with specified elements and in a non-null model.

(c) *public JList (object [ ] listData):* This constructor creates a **JList**, which displays elements of the array "ListData".

The above mentioned constructors help the user to create a **JList** component in different forms.

 Example:

Program to illustrate the list box component in Java.

```java
import javax.swing.*;
import javax.swing.event.ListSelectionEvent;
import javax.swing.event.ListSelectionListener;
import java.awt.*;
import java.applet.*;
import java.awt.event.*;
public class JListDemo extends JApplet
{
  JList list;
  String[ ] listColorNames = { "black", "blue", "pink", "gray","white" };
  Color[ ] listColorValues = { Color.BLACK, Color.BLUE,
            Color.PINK,Color.GRAY, Color.WHITE};
  Container contentpane;
  public JListDemo( )
  {
    super( );
    contentpane = getContentPane( );
    contentpane.setLayout(new FlowLayout( ));
    list = new JList(listColorNames);
    list.setSelectedIndex(0);
    list.setSelectionMode(ListSelectionModel.SINGLE_SELECTION);
    contentpane.add(new JScrollPane(list));
```

```
        list.addListSelectionListener(new ListSelectionListener( )
{
 public void valueChanged(ListSelectionEvent e)
 {
   contentpane.setBackground(listColorValues[list.getSelectedIndex( )]);
 }
 }
  setSize(200, 200);
  setVisible(true);
 }
 public static void main(String[ ] args)
 {
 JListDemo test = new JListDemo( );
 }
}
```

HTML Coding:

```
<html>
<head></head>
<title></title>
<body>
<p>
<applet code="JListDemo.class" width=225 height=200>
</applet>
</p>
</body>
</html>
```

Output:

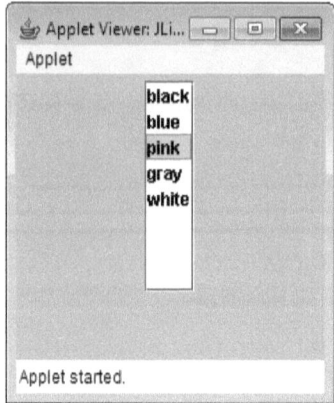

In this example,

1.   First, a class **JListDemo** is created, which extends the class JApplet.

2.   In this class,

   (a)   A JList object **list** is declared, the **listColorNames** and **listColorValues** objects are defined, the container **contentpane** is declared, and the JListDemo method is called.

   (b)   In this method,

      (i)   The getContentPane( ) method is called to get the content pane; this content pane is linked to the container **contentPane**.

      (ii)   A new FlowLayout is created while setting the layout of the content pane with the help of setLayout( ) method.

     (iii)   Then, a new JList is created with the string listColorNames, and is linked to the list.

     (iv)   The first element of the list is set as selected using the setSelectedIndex method.

      (v)   The selection mode of the list is set to SINGLE_SELECTION using the setSelectionMode method.

     (vi)   The list is then added to the content pane, by using the add( ) method.

    (vii)   A list selection listener method ListSelectionListener is added to the list using the method addListSelectionListener. The ListSelectionListener method will now be called when a different entry in the list is selected.

    (viii)   The ListSelectionListener method calls the valueChanged   method, passing the ListSelectionEvent as a parameter.

     (ix)   The valueChanged method sets the background color of the container to the color mentioned in the selected index of the object listColorValues.

     (x)   The setSize and setVisible methods are called to set the size and make the container visible.

(xi) Finally, the JListDemo method is invoked from the main program.

Note: Finally, the code is executed in the command prompt. For acquiring the output, an HTML code is written. The result is shown as an applet window.

Task   Compile a list of constructors used in the swing class, along with their description.

Menus

Menu is a part of GUI, which is used to display a list of items indicating various options that can be used by a user. The user can click on any option and a sub-menu, if present, appears on the screen. Every menu item has an action associated with it. Similarly, the Swing menu contains menubar, menuitems and menu. Menubar is regarded as the root of all menus and menuitems.

A Swing menu has many components, namely, JMenuBar, JPopUpMenu, JAbstractButton and JSeparator. The JAbstractButton further has one component, namely, JMenuItem component. The JMenuItem component contains other components like JMenu, JCheckBoxMenuItem and JRadioButtonMenuItem.

1. *JMenuBar:* **JMenuBar** is defined as a swing component, which can be added to a container only through JFrame, JWindow and JInternalRootFrame's root pane.. **JMenuBar** contains various JMenus, and every **JMenu** depicts the string within the **JMenubar**. When the user clicks on any string, the menu associated with it appears on the screen displaying different menu items.

Notes   A JFrame is a swing component, which is used as a support for the swing component architecture. The JRootPane is contained in a JFrame. The JRootPane is the only child of JFrame. The content pane, which the root pane provides must include all the non-menu components that are displayed by the JFrame. A container, which can be displayed on a user's desktop, is called as JWindow. The title bar, window-management buttons, or other trimmings which are associated with a JFrame, are not present in a JWindow.

Example: Program to illustrate the JMenuBar component in Java.

```
import java.awt.*;

import javax.swing.*;

public class MenuBar extends JApplet
{
  public void init( )
  {
    JMenuBar m = new JMenuBar( );

    JMenu fileMenu = new JMenu("Display");

    JMenu pullRightMenu = new JMenu("pull right");

    fileMenu.add("welcome");

    //fileMenu.addSeparator;

    fileMenu.add(pullRightMenu);
```

```
fileMenu.add("Exit");
pullRightMenu.add(new JCheckBoxMenuItem("Hi"));
pullRightMenu.add(new JCheckBoxMenuItem("Hello"));
pullRightMenu.add(new JCheckBoxMenuItem("How are you?"));
m.add(fileMenu);
setJMenuBar(m);
 }
}
```

HTML Coding:

```
<html>
<head></head>
<title></title>
<body>
<p>
<applet code="MenuBar.class" width=225 height=200>
</applet>
</p>
</body>
</html>
```

Output:

In this example,

1. First, a class **MenuBar** is created, which extends the class JApplet.

2. In this class, the **init( )** method is called.

In this method,

(a) A new JMenuBar is created, and is linked to the JMenuBar **m**.

(b) The string **Display** is linked to JMenuFileMenu using the new keyword.

(c) The string **pullright** is linked to JMenu pullRightMenu using the new
keyword.

(d) The **pullRightMenu** is added to the fileMenu using the add( ) method.

(e) The string **Exit** is then added to the fileMenu, by using the add( ) method.

(f) The string **Hi** is added to the **pullRightMenu** using the add( ) method and linked to JCheckBoxMenuItem using the keyword new.

(g) The string **Hello** is added to the pullRightMenu using the add( ) method and linked to JCheckBoxMenuItem using the keyword new.

(h) The string **How are you?** is added to the **pullRightMenu** using the add( ) method and linked to JCheckBoxMenuItem using the keyword new.

(i) The fileMenu is then added to **m**, by using the add( ) method.

(j) The JMenuBar is added to the container.

Note: Finally, the code is executed in the command prompt. For acquiring the output, an HTML code is written. The result is shown as an applet window with the text area and text field in the applet.

2. *JMenu:* **JMenu** is also a Swing menu component. JMenu is a component of the JMenuItem and comes under JMenuBar. This component serves two purposes.

(a) First, when the user clicks it, a text string appears on the screen.

(b) Second, when the user clicks on the string, a popup menu appears.

JMenu contains JMenuItem, JCheckBoxItem, JRadioButtonItem and JSeparator.

3. *JPopUpMenu:* The **JPopUpMenu** is another Swing menu component. The **JPopUpMenu** component is used to display the expanded form of the menu. This component can be used in two ways:

(a) Firstly, the **pull-right** menu appears on the screen, when the user clicks a menu item.

(b) Secondly, it can be used as a shortcut menu, which gets activated when a user right clicks on the mouse.

A **JPopUpMenu** can be created using two constructors, which are:

(a) *public JPopUpMenu( ):* This constructor creates a JPopUpMenu.

(b) *public JPopupMenu(String label):* This constructor creates a JPopUpMenu with a specified title.

The constructors mentioned above help the user to create a JPopUpMenu component, with a title or without a title.

4. *JMenuItem:* **JMenuItem** is defined as a Swing menu component, which appears on the screen in a string format with an icon. A menu is contained in the **Jmenultem** component. When a user clicks and releases the mouse on a **JMenuItem**, the menu which contains the item disappears. Then, a dialog box appears, which is displayed for that menu item. The attributes of a JMenuItem like font, color, background, and border can be changed.

5. *JCheckBoxMenuItem:* The **JCheckBoxMenuItem**, a Swing menu component, consists of checkboxes as the menu items. These checkboxes can be created using the **JCheckBox** class. The attributes of a **JCheckBoxMenuItem** like the appearance, background color and the icon associated with the checkbox, can be changed.

6. *JRadioButtonMenuItem:* The **JRadioButtonMenuItem** is a swing menu component, which can be selected only one at a time. It is similar to the checkbox component. If a user clicks on the radio button, then the following possibilities occur:

   (a) If the item is selected before, then the state does not change.

   (b) If the item was not selected initially, the selected item is deselected and the clicked item is selected.

   If a **JRadioButtonMenuItem** changes its state, then the **JRadioButtonMenuItem** creates an **ItemEvent** and distributes it to the registered **ItemListeners**.

Lab Exercise

1. Write a program to illustrate the concept of checkboxes in Java. Consider four checkboxes with different names.

2. Write a program to illustrate the usage of buttons with background color as pink.

3. Write a program to illustrate a menu in Java. The words **file** and **help** must appear on the menu screen.

Case Study

Success Story of ProSeco Project

ProSeco was intended for the TV-broadcasting planning. It involved all the planning tasks hat could possibly arise in a broadcasting industry. It involved tasks from strategic rogram planning, through broadcasting planning up to broadcasting operations.

Pr....... .... a very large project that was developed by two teams. These two teams included both the in-house as well as the offshore teams. This project tried a new model of interaction between the two development teams. It was decided that the in-house team would design the project, while the offshore team would perform programming.

The in-house team comprised system analysts, an architect and a designer. The designer was responsible for all the communication that took place with the offshore team. It was the designer who transformed the design team requirements into technical solutions, documented and then passed on the data to the team. Initially, the technical solution had been specified at a low level, that is, in terms of classes and methods. However, as the offshore team gained experience and knowledge of the system, tasks description became less detailed. Although the volume of communication decreased, the efficiency of work of the team grew.

The team specialists had to study the original components and then employ them during the development of the complicated server application. One of the major contributing factors to the success of this project was the implementation of Juice, which is a Swing extension envelope intended for the development of XML-driven GUI.

Questions:

1. What is the purpose of the ProSeco Project?

2. How was this project different from other projects in terms of interaction among the team members?

3. Implementation of which tool led to the success of this project?

Source: http://www.soft-product.com/index.php?id=130

13.3 Summary

- The concept of Swings in Java uses AWT's component. The architecture of the Swing components modifies the program's appearance and behavior.

- AWT package **javax.swing** contains all the Swing components; namely: JComponent , JButton, JCheckBox, JFileChooser, JTextField, JFrame, JPanel, JApplet, JOptionPane, JDialog and JColorChooser.

- **JLabel** is used to display an area for a short text string or an image, or both, and they do not respond to input events.

- The text components are Swing components, which are used to display text, and allow the user to edit the text.

- Six text components are present in Swing namely, JTextField, JFormattedTextField, JPasswordField, JTextArea, JTextPane and JEditorPane.

- **Checkbox** is a Swing component, which is used to select or deselect an item in a list with a single click.

- A radio button is a Swing component, which is similar to the checkbox component. But in a group of radio buttons, only one radio button can be selected per click, unlike check boxes.

- **Menus** are a part of Swing components and are used to display a list of items, which indicates various options that can be used by a user.

- In Swings, menu consists of MenuBar, MenuItems and Menus. The MenuBar is known as the root of all menus and menu items.

13.4 Keywords

GUI: Graphical User Interface

java.awt: A user interface toolkit called the Abstract Windowing Toolkit, or the AWT, which is provided by the Java programming language class library.

java.awt.event: A package in Java that provides interfaces and classes, which are used to deal with various events that are fired by the AWT components.

javax.swing: A package in Java that provides a collection of lightweight components, that is, the components working similarly on all platforms.

JFC: Java Foundation Class

JSeparator: A Swing component that provides a general purpose component, which is used to implement divider lines that are generally used as a divider between menu items. This divider divides these menu items into logical groups.

Tabbed Panes: A stack of components in selectable layers

13.5 Self Assessment

1. State whether the following statements are true or false:

    (a) Although the Swing components are not lightweight, they provide a better user interface.

    (b) Swing provides the multiple platform independent API interfaces.

    (c) Swing library is used to create graphical user interfaces in java.

    (d) The **JMenuItem** is defined as a component, which appears on the screen in a button format with an icon.

    (e) Menus are used to display a list of items, which indicates various operations that can be used by a user.

    (f) Checkbox is defined as a component, which combines a button or editable field and a drop-down list.

2. Fill in the blanks:

    (a) _____provides the multiple platform independent API interfaces.

    (b) Swing is a _package.

    (c) Swings have _____public packages.

    (d) JtextArea is known as a lightweight component for working with _____ .

    (e) JTextPane can exhibit _____documents.

3. Select the suitable choice for every question.

    (a) Which of the following components extends and matches the Frame class in the AWT package?

        (i) JFrame

        (ii) JPanel

        (iii) JApplet

        (iv) JDialog

    (b) JCheckBox and JCheckBoxMenuItem are inherited from which component's class?

        (i) ComboBoxes

        (ii) Lists

        (iii) Menu

        (iv) Button

    (c) Identify which of the following constructors is not used to create the JList component?

        (i) public JList( )

        (ii) public JList (ListModel dataModel)

        (iii) public JList (ListSelectionListener)

        (iv) public JList (object [ ] listData)

    (d) Which of the following components does the JCheckBox class support?

        (i) Combo box

        (ii) Check box

        (iii) List

        (iv) Radio button

    (e) Which of the following objects are created by using the **JRadioButton( )** and **JRadioButton(Icon icon)** constructors?

        (i) JRadioButton

        (ii) JButton

        (iii) JTextBox

        (iv) JCheckBox

13.6 Review Questions

1. "A Swing is the graphical user interface library of Java." Justify this statement.

2. "The Java Abstract Window Toolkit (AWT) provides a platform-specific code." How is a swing different from an AWT?

3. "The **javax.swing.SwingWorker** class is the generic solution to the issues of updating the GUI from worker threads and giving users the ability to control the background tasks." Discuss.

4. "**JComponent** class is derived from the container class of AWT". Elaborate.

5. "Swing packages are completely written in Java, whereas AWT are not completely written in Java." Comment.

6. "Frames can be closed by just clicking on the close button." Discuss.

7. "The label can be aligned by setting the label's contents using the vertical and horizontal alignment." Justify this statement with the help of an example.

8. "Swing buttons are used to display both text and an image." Discuss.

9. "A **JRadioButton** object can be created using many constructors." Discuss these constructors.

10. "A JFrame is a swing component, which is used as a support for the swing component architecture." Do you agree? Justify.

11. "Swing checkboxes have common button characteristics." Explain.

12. "A **ComboBox** contains three components." Discuss these components.

Answers: Self Assessment

1.   (a) False   (b) True   (c) True   (d) False

    (e) True   (f) False

2.   (a) Swing   (b) Java   (c) 18   (d) Text

    (e) HTML

3.   (a) JFrame   (b) Button   (c) public JList(ListSelectionListener)

    (d) CheckBox   (e) JRadioButton

13.7 Further Readings

Books   Er. R. Kabilan, (2009), Secrets of JAVA, Firewall Media

             Kim Topley , (2000), Core Swing: advanced programming, Prentice Hall PTR

Online link   http://java.comsci.us/examples/swing/JTextArea.html

             http://www.javabeginner.com/java-swing/java-swing-tutorial

             http://zetcode.com/tutorials/javaswingtutorial/basicswingcomponentsII

             http://www.devx.com/tips/Tip/12812

             http://www.beginner-java-tutorial.com/jbutton.html

Unit 14: Event Handling

Objectives

After studying this unit, you will be able to:

- Describe the event delegation model
- Explain the event classes
- Illustrate the use of various event classes

Introduction

As applets are important elements of Java programming, similarly the concept of event handling is the key to programming in applets. An event is an object which elaborates the change of a state in a source. Since Java has the mechanism to handle events occurring in any program or application, Java is sometimes called as **event-driven** programming language.

Sometimes events occur in programs which must be handled with utmost care for the program's proper execution. It is necessary for many programs to be responsive to the commands given by the user. For this purpose, Java programs depend on events that specify user actions. Just like exception handlers, event handlers are also inherently called whenever something (an event) occurs. These events are generated by external actions, such as interactions of the user through a GUI.

Did you know?   Event handling helps in making web applications more dynamic and interactive.

The concept of event handling covers many important elements, such as events, events listeners, event source, and so on. All these elements together make event handling successful.

14.1 Meaning of Event Handling

Event handling is an important aspect that relates to applets in Java. An event is known as the core of applet programming. The **java.awt.event** package comprises these events. The events to which the applets respond are generated by the user. These events are passed to the applets in several ways.

In a program, two methods **action( )** or **handleEvent( )** are used for catching and processing GUI events. But to do so, it must subclass GUI components and override the **action( )** or **handleEvent( )** methods. If this method returns a **true** value, then the event is not handled further. In case, a **true** value is not returned, then the event is spread sequentially up the GUI hierarchy until it reaches the root of the hierarchy. This model returning **true** or **false** values results in providing the user two choices to give a structure to the event handling code:

1. The individual component can be sub-classed to explicitly handle its target events.

2. Every event of the GUI hierarchy can be handled by a particular container. Then, the container's overridden methods like the **action( )** or **handleEvent( )** should contain a   complex conditional statement to process the events.

14.2 The Event Delegation Model

As exceptions are handled in Java, so are events. In Java, event models are used for handling events. One such model is the **event delegation model**.

The event delegation model is considered as a modern approach to handle events. This model introduced the concept of listeners. It defines standard and consistent mechanism to generate and process events. The bases of an event model are event source and event listeners. An event source refers to an object generating these events, whereas an event listener refers to the object interested in receiving these events. In this system, the listener just waits until it receives an event. Once an event is received, the listener processes the event and then returns.

The advantage of this design is that the application logic that processes events is cleanly separated from the user interface logic that generates the events, wherein delegation of the event processing to a unique segment of code can be done by a user interface element.

In the event delegation model, the listeners must register with a source for receiving an event notification. This is advantageous where the notifications are sent merely to listeners, who are willing to receive them. This way of handling events is more effective. The basic idea behind creating such a model is to provide a robust framework to the Java programs.

Some of the basic goals for creating the delegation model are as follows:

- The delegation model is simple and easy to learn.

- The model supports a separation between application and GUI code.

- The model eases the creation of robust event handling code.

- The delegation model is flexible and enables varied application models for event flow and propagation.

- This model supports the backward binary compatibility with the old model.

The event delegation model can be used easily. Steps to use this model are:

1. Execute the appropriate interface in the listener so that it will receive the type of event desired.

2. Implement code to register and unregister listener as a recipient for the event notification.

Notes

In Java, processing of events without the help of delegation event model is permitted. An AWT component is extended to do so.

Example: Program to illustrate the concept of mouse event handlers.

```
import java.awt.*;
import java.awt.event.*;
import java.applet.*;

/*
<applet code= "MouseEvents" width=200 height=100 >
</applet>
*/

public class MouseEvents extends Applet
implements MouseListener, MouseMotionListener
{
String msg = "" ;
int mouseX = 0, mouseY = 0; //coordinates of mouse
public void init( )
{
addMouseListener (this);
addMouseMotionListener(this);
}

//Handle mouse clicked.
public void mouseClicked(MouseEvent me)
{
//save coordinates
mouseX = 0;
mouseY = 10;
msg = "Mouse clicked.";
repaint( );
}

//Handle mouse entered.
public void mouseEntered(MouseEvent me)
{
//save coordinates
```

```
mouseX = 0;
mouseY = 10;
msg = "Mouse entered.";
repaint( );
}

//Handle mouse exited.
public void mouseExited(MouseEvent me)
{
//save coordinates
mouseX = 0;
mouseY = 10;
msg = "Mouse exited.";
repaint( );
}

//Handle button pressed.
public void mousePressed(MouseEvent me)
{
//save coordinates
mouseX = me.getX( );
mouseY = me.getY( );
msg = "Down.";
repaint( );
}

//Handle button released.
public void mouseReleased(MouseEvent me)
{
//save coordinates
mouseX = me.getX( );
mouseY = me.getY( );
msg = "Up. ";
repaint( );
}

//Handle mouse dragged.
public void mouseDragged(MouseEvent me)
{
//save coordinates
mouseX = me.getX( );
mouseY = me.getY( );
msg = "*. ";
showStatus("Dragging  mouse at  " + mouseX +  " , " + mouseY);
repaint( );
}

//Handle mouse moved.
public void mouseMoved(MouseEvent me)
{
//show status
showStatus("  Moving mouse at " + me.getX( ) + ", " + me.getY( ) );
}

//Display msg in applet window at current X, Y location.
public void paint (Graphic g)
{
```

g.drawString(msg, mouseX, mouseY);
}
}

Output:

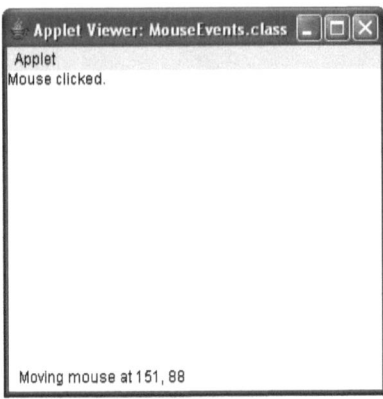

(We get the above output when we click the mouse button inside the applet)

In this example,

1. First, the **MouseEvents** class extends the Applet class and implements both the MouseListener and MouseMotionListener interfaces. These two interfaces comprise methods used for receiving and processing different types of mouse events. Notice that the applet is both the source and the listener for events. This is a common situation for applets.

2. In the init( ) method of the **MouseEvents** class, the applet registers itself as a listener for mouse events. This is done by using addMouseListener( ) and addMouseMotionListener( ), which are the members of Component class. Their general forms are:

   void addMouseListener(MouseListener ml)
   void addMouseMotionListener(MouseMotionListener mml)
   Here, **ml** is a reference to the object receiving mouse event, and **mml** is reference to the object receiving mouse motion events. In the above program, the same object is used for both.

3. Then, the applet implements all the methods that are specified by the two interfaces, MouseListener and MouseMotionListener. These are event handlers which handles different types of mouse events. Every method handles the events associated with them and finally gives the appropriate output.

Task   Demonstrate some virtual key codes.

14.2.1 Events

In the event delegation model, an event refers to an object, which elaborates the change of a state in a source. Some of the activities that cause the generation of events are entering a character via the keyboard, a button press, opting for a list item, and mouse click.

Sometimes events may also arise because of interactions with a user interface. You are free to define events that are appropriate to your application.

 Example: An event may be generated when a timer terminates, a counter exceeds a value, software or hardware failure occurs, or an operation is completed.

The following steps are followed to set up the processing of events:

1. First, the GUI component is associated a listener object class with the component by calling a
    addXListener method.

2. Then, the listener object used is defined. The object must implement the corresponding interface whose name is of type **EventListener**.

3. Finally, all the methods must be defined by the object. This must be done in the interface the object is implementing.

14.2.2 Event Sources

In the event delegation model, a source event occurs when there is a modification in the object's internal state, in some way. Event sources may produce more than one type of events. Sources may generate more than one type of event. For the listeners to receive notifications about a specific event, it must register with the source. Each type of event has its own registration method. The general form of event source:

> public void addTypeListener (*Type*Listener *el*)

Here, the **Type** is the name of the event, and **el** is a reference to the event listener.

 Example: The **addKeyListener( )** method is used for registering keyboard event listeners, whereas the **addMouseMotionListener( )** method is used for registering mouse motion listeners.

When an event takes place, all registered listeners are notified and receive a copy of the event object.

 The process of notifying the registered listeners and providing them a copy of the event object on the occurrence of an event is called as multicasting an event.

Some of the event sources permit the registration of only one listener. The general form of these event sources is:

> public void addTypeListener (*Type*Listener *el*)
>
> throws java.util.TooManyListenerException

Here, **Type** is the name of the event, and **el** is a reference to the event listener.

 Example: The **removeKeyListener( )** method is called to remove a keyboard listener.

Sometimes, on the occurrence of any event, only the registered user is notified. This type of casting is called as unicasting the event.

A source must also provide a method that allows a listener to un-register an interest in a specific type of event. The general form of this event source is:

> public void removeTypeListener(*Type*Listener *el*)

Here, **Type** is the event name, and **el** is a reference to the event listener, and the **removeKeyListener( )** method is called to remove a keyboard listener.

The methods that add or remove listeners are given by the source that generates events. For example, the Component class provides methods to add and remove keyboard and mouse event listeners.

14.2.3 Event Listeners

Event Listeners are objects and are used to handle a particular task of a component. The Listener implements the interface which contains event-handling code for a particular component. Event listeners have two main requirements.

1. First, the event must be registered with one or more sources for receiving notifications about certain events on their occurrence.

2. Second, the event must execute methods for receiving and processing these notifications.

The methods that receive and route events are defined in a set of interfaces found in the **java.awt.event** package.

 *Example:* The **MouseMotionListener** interface defines two methods to receive notifications when the mouse is dragged or moved. Any object may receive and process one or both of these events if it provides an execution of this interface.

The listeners are registered by an event source, and to handle an **ActionEvent**, a **Button** may register an object, by calling **addActionListener**. This object would then implement the listener interface corresponding to **ActionEvent**, which is **ActionListener**.

Table 14.1 lists the commonly used listener interfaces and provides a brief description of the methods that they define.

Table 14.1: Commonly Used Event Listener Interfaces	
Interface	**Description**
ActionListener	Specifies one method for receiving action events.
AdjustmentListener	Specifies one method for receiving adjustment events.
ItemListener	Specifies one method for checking when the state of an item gets modified.
TextListener	Specifies one method to find out when a text value changes.
ComponentListener	Specifies four methods to find out when a component is moved, hidden, shown or resized.
ContainerListener	Specifies two methods to find out when a component is added to or removed from a container.
FocusListener	Specifies two methods to find out when a component gains or loses keyboard focus.
KeyListener	Specifies three methods to find out when a key is pressed, released or typed.
MouseListener	Specifies five methods to find out when the mouse is clicked, enters a component, leaves a component, is pressed or is released.
MouseMotionListener	Specifies two methods to find out when the mouse is dragged or moved.
MouseWheelListener	Specifies one method to find out when the mouse wheel is moved.
WindowListener	Specifies seven methods to find out when a window is activated, closed, deactivated, iconified, deiconified, opened or quit.
WindowFocusListener	Specifies two methods to find out when a window gains or loses input focus.

The ActionListener Interface

This interface specifies the **actionPerformed( )** method that is invoked when an action event occurs. Syntax of the ActionListener interface:

  void actionPerformed(ActionEvent ae)

The AdjustmentEventListener Interface

This interface specifies the **adjustmentValueChanged( )** method that is invoked when an adjustment event occurs. Syntax of the AdjustmentEventListener interface:

void adjustmentValueChanged(AdjustmentEvent ae)

The ComponentListener Interface

The interface specifies four methods that are invoked when a component is resized, moved, shown or hidden.

1. *void componentResized(ComponentEvent ce):* This method is invoked when a component is resized.

2. *void componentMoved (ComponentEvent ce):* This method is invoked when a component is moved.

3. *void componentShown(ComponentEvent ce):* This method is invoked when a component is shown.

4. *void componentHidden(ComponentEvent ce):* This method is invoked when a component is hidden.

The ContainerListener Interface

This interface contains two methods.

1. *componentAdded( ):* This method is called when a component's addition is made to a container. Its syntax is:

    void componentAdded(ContainerEvent ce)

2. *componentRemoved( ):* This method is called when a component's removal is done from a container. Its syntax is:

    void componentRemoved(ContainerEvent ce)

The FocusListener Interface

This interface specifies two methods.

1. *focusGained( ):* This method is called when a component obtains a keyboard focus. Its syntax is:

    void focusGained(FocusEvent fe)

2. *focusLost( ):* This method is called when a component loses keyboard focus. Its general form is as follows:

    void focusLost(FocusEvent fe)

The ItemListener Interface

This interface specifies the **itemStateChanged( )** method that is invoked when the state of an item changes. Its syntax is:

    void itemStateChanged(ItemEvent ie)

The KeyListener Interface

This interface specifies three methods.

1. *keyPressed( ):* This method is called when a key is pressed. Its syntax is:

    void keyPressed(KeyEvent ke)

2. *keyReleased( ):* This method is called when a key is released. Its syntax is:

    void keyReleased(KeyEvent ke)

3. *keyTyped( ):* This method is called on the entry of a character. Its syntax is:

    void keyTyped(KeyEvent ke)

Example: If a user presses and releases the **A** Key, three events are generated in a sequence:
Key pressed
Key typed
Key released
If a user presses and release the **HOME** key, two key events are generated in a sequence:
Key pressed
Key released

The MouseListener Interface

This interface specifies five methods.

1. *mouseClicked( ):* This method is called when the mouse is pressed and at the same time is released. Its syntax is:

    void mouseClicked(MouseEvent me)

2. *mouseEntered( ):* This method is invoked when an entry is made by the mouse into a component. Its syntax is:

    void mouseEntered(MouseEvent me)

3. *mouseExited( ):* This method is called when the component is left by the mouse. Its syntax is:

    void mouseExited(MouseEvent me)

4. *mousePressed( ):* The MouseListener interface calls this method when the mouse is pressed. Its syntax is:

    void mousePressed(MouseEvent me)

5. *mouseReleased( ):* The MouseListener interface calls this method when the mouse is released. Its syntax is:

    void mouseReleased(MouseEvent me)

The MouseMotionListener Interface

This interface specifies two methods.

1. *mouseDragged( ):* This method is invoked more than once whenever the mouse is dragged. Its syntax is:

    void mouseDragged(MouseEvent me)

2. *mouseMoved( ):* This method is called more once whenever the mouse is moved. Its syntax is:

    void mouseMoved(MouseEvent me)

The MouseWheelListener Interface

This interface specifies the **mouseWheelMoved( )** method that is invoked when the mouse wheel is moved. Its syntax is:

void mouseWheelMoved(MouseWheelEvent mwe)

The TextListener Interface

This interface specifies the **textChanged( )** method that is called when a change occurs in a text area or a text field. Its syntax is:

void textChanged(TextEvent te)

The WindowFocusListener Interface

This interface specifies two methods.

1. *windowGainedFocus( ):* This method is called when a window gains input focus. Its syntax is:

   void windowGainedFocus(WindowEvent we)

2. *windowLostFocus( ):* This method is called when a window loses input focus. Its syntax is:

   void windowLostFocus(WindowEvent we)

The WindowListener Interface

This interface specifies seven methods.

1. *windowActivated( ):* This method is invoked on the activation of a window. Its syntax is:

   void windowActivated(WindowEvent we)

2. *windowDeactivated( ):* This method is invoked on the deactivation of a window. Its syntax is:

   void windowDeactivated(WindowEvent we)

3. *windowIconified( ):* This method is invoked whenever a window is iconified. Its syntax is:

   void windowIconified(WindowEvent we)

4. *windowDeiconified( ):* This method is invoked whenever a window is deiconified. Its syntax is:

   void windowDeiconified(WindowEvent we)

5. *windowOpened( ):* This method is invoked whenever a window is opened. Its syntax is:

   void windowOpened(WindowEvent we)

6. *windowClosed( ):* This method is invoked whenever a window is closed. Its syntax is:

   void windowClosed(WindowEvent we)

7. *windowClosing( ):* This method is invoked whenever a window is being closed. Its syntax is:

   void windowClosing(WindowEvent we)

All these interfaces are very important in the event handling, as they are some of the vital elements of the event delegation model.

14.3 Event Classes

In Java, an event class is a class of events. An event is generated when the user interacts with a GUI application.

 Example: Few examples of user events are clicking a button, selecting an item or closing a window.

Events are represented as Objects in Java. The **java.util.EventObject** is the super class of all event classes. At the root of the Java event hierarchy is **EventObject**, which is available in **java.util** package. It is the super class for all events.

Constructor Used to Create an Event Class Object:

   EventObject(Object *src*)

Here, **EventObject** is the name of the object of the **event** class, **src** is the object that produces this event. An event object contains two methods, **getSource( )** and **toString( ).**

1. *getSource( ):* The **getSource( )** method is used to get the event source. Its general form is:

   Object getSource( )

2. *toString( ):* The **toString( )** method is used to get the string equivalent of the event. Its general form is:

    Object toString ( )

The EventObject class has a subclass AWTEvent, which is defined within the java.awt package. This subclass is the super class of all AWT-based events used by the delegation event model. Its **getID( )** method is used to determine the type of the event. Syntax of this method:

    int getID( )

Here, **int** refers to the integer data type and **getID( )** is the method name.

The subclasses of **AWT Event** class can be categorized into two groups.

1. *Semantic events:* Events, which directly correspond to high-level user interactions with any GUI component.

   Example: Clicking of a button is a semantic event.

Some of the semantic event classes are:

(a)   ActionEvent

(b)   AdjustmentEvent

(c)   ItemEvent

(d)   TextEvent

2. *Low-level events:* Multiple low-level events may be produced for all the high-level user events. Some of the low level event classes are:

(a)   ComponentEvent

(b)   ContainerEvent

(c)

      FocusEvent

(d)   InputEvent

(e)   KeyEvent

(f)   MouseEvent

(g)   MouseWheelEvent

(h)   PaintEvent

(i)   WindowEvent

The main event classes are listed in table 14.2.

Table 14.2: Main Event Classes in java.awt.event Package	
Event Class	**Description**
ActionEvent	Produced whenever any button is pressed, an item is double-clicked in a list item or an item is selected in a menu.
AdjustmentEvent	Produced on the manipulation of a scroll bar.
ComponentEvent	Produced whenever any component is hidden, resized, moved, or becomes visible.
FocusEvent	Produced when a component gains or loses keyboard focus.
ItemEvent	Produced whenever any check box or list item is clicked, a choice is made, or an item in a checkable menu is selected or deselected.
KeyEvent	Produced when the keyboard receives the input.
MouseEvent	Produced whenever the mouse is dragged, moved, clicked, pressed or released, enters or exits a component.
MouseWheelEvent	Produced whenever the mouse wheel is moved.
TextEvent	Produced whenever the value of a text area or text field is modified.
WindowEvent	Produced whenever a window is activated, closed, deactivated, deiconified, iconified, opened or quit.

14.3.1 The ActionEvent Class

We know that the **ActionEvent** class is a semantic event class. The **ActionEvent** takes place, whenever any button is pressed, an item in a list is clicked twice, or an item is selected in a menu. The **ActionEvent** class defines four integer constants that can be used to identify an action event, and modifiers associated with an action event, ALT_MASK, CTRL_MASK, META_MASK and SHIFT_MASK. Apart from these constants, there is one more integer constant, ACTION_PERFORMED, which can be used in the action event's identification.

Constructors of the ActionEvent Class

1. ActionEvent(Object src, int type, String cmd)

2. ActionEvent(Object src, int type, String cmd, int modifiers)

3. ActionEvent(Object src, int type, long when, String cmd)

In these three constructors, **src** is a reference to the object by which this event (ActionEvent) was generated. The type of the event is specified by **type** and its command string is specified by **cmd**. The argument modifier indicates which modifier keys (CTRL, ALT, META, and SHIFT) were pressed, when the event was generated; the **when** parameter is used for specifying when the event occurred.

Methods Used in the ActionEvent Class

1. *getActionCommand( ):* This method can be used to obtain the command name for invoking **ActionEvent** object. Syntax of this method:

    String getActionCommand( )

 Example: When a button is pushed, an action event is created that has a command name similar to the label on that button.

2. *getModifiers( ):* This method is used to return/obtain a value that shows which modifier keys (CTRL, ALT, META and SHIFT) were pressed, when the event was created. Syntax of this method:

    int getModifiers( )

3. *getWhen( ):* This method is used to return the time when the event occurred. This is called event's **timestamp**. Syntax of this method:

    long getWhen( )

14.3.2 The AdjustmentEvent Class

Just like ActionEvent class, the AdjustmentEvent class is also a semantic event class, wherein a scroll bar generates an AdjustmentEvent. The AdjustmentEvent class defines the constants of integer type that can be used to identify these events. These constants are depicted in the table 14.3.

Table 14.3: Constants and their Meaning

Constants	Meaning
BLOCK_DECREMENT	The value of the scrollbar decreases, when the user clicks inside it.
BLOCK_INCREMENT	The value of the scrollbar increases, when the user clicks inside it.
TRACK	The slider is dragged.
UNIT_DECREMENT	The button at the end of the scroll bar is clicked for decrementing its value.
UNIT_INCREMENT	The button at the end of the scroll bar wasis clicked for incrementing its value.

In addition to the above constants, there is one more integer constant, ADJUSTMENT_VALUE_CHANGED, which specifies that a change has occurred.

Constructor of the AdjustmentEvent Class

    AdjustmentEvent(Adjustable *src*, int *id*, int *type*, int *data*)

In this constructor, **src** is a reference to the object that generates the event. The **id** specifies the event. The type of the adjustment is specified by **type**, and its associated data is **data**.

Methods Used in AdjustmentEvent Class

1. ***getAdjustable( ):*** This method is used to get the object that generates the event. The syntax of this method is:

    Adjustable getAdjustable( )

2. ***getAdjustableType( ):*** This method is used to get the type of the adjustment event. It returns one of the constants defined by the AdjustmentEvent. The syntax of this method is:

    int getAdjustment Type( )

3. ***getValue( ):*** This method is used to get the amount of the adjustment. The syntax of this method is:

    int getValue( )

 Example: When a scroll bar is manipulated, the method returns the value represented by that change.

14.3.3 The ComponentEvent Class

The **ComponentEvent** class is a low-level event class, wherein a ComponentEvent is produced when a component's size, position or visibility is modified. The ComponentEvent class specifies constants of integer type that can be used for identifying these events. These constants are shown in the table 14.4.

Table 14. 4: Constants and their Meaning	
COMPONENT_HIDDEN	The component was hidden.
COMPONENT_MOVED	The component was moved.
COMPONENT_RESIZED	The component was resized.
COMPONENT_SHOWN	The component became visible.

Constructor of the ComponentEvent Class

ComponentEvent(Component *src*, int *type*)

In this constructor, **src** is a reference to the object that generates this event. The type of the event is specified in **type**.

ComponentEvent class is the parent class of the ContainerEvent class, FocusEvent class, KeyEvent class, MouseEvent class and WindowEvent class either directly or indirectly.

Method Used in the ComponentEvent Class

getComponent( ): This method is used to obtain the component that generates the event. Syntax of this method:

Component getContainer( )

14.3.4 The ContainerEvent Class

The **ContainerEvent** class is a low-level event class, wherein a ContainerEvent is created, when a component is added to or removed from a container. The ContainerEvent class defines **int** constants that can be used to identify the events of the class as:

1. COMPONENT_ADDED
2. COMPONENT_REMOVED

These integer constants specify that a component has been added to or removed from the container. The ComponentEvent class is the parent class of the ContainerEvent class.

Constructor of the ContainerEvent Class

ContainerEvent(Component *src*, int *type*, Component *comp*)

In this constructor, **src** is a reference to the container by which this event was produced. The type of the event is passed in the parameter **type**, and the component that has been added to or removed from the container is **comp**.

Methods Used in the ContainerEvent Class

1. ***getContainer( ):*** This method is used to get a reference to the container that generated this event. The syntax of this method is:

    Container getContainer( )

2. ***getChild( ):*** This method is used to get a reference to the component's reference that was added to or removed from the container. The syntax of this method is:

    Component getChild( )

14.3.5   The FocusEvent Class

The **FocusEvent** class is also a low-level event class, wherein a FocusEvent is generated when a component gains or loses input focus. Integer constants such as FOCUS_GAINED and FOCUS_LOST are used to identify these events. **FocusEvent** class is a subclass of **ComponentEvent** class.

Constructors of the FocusEvent Class

1. FocusEvent(Component *src*, int *type*)

2. FocusEvent(Component *src*, int *type*, Boolean *temporary Flag*)

3. FocusEvent(Component *src*, int *type*, Boolean *temporary Flag*, Component *other*)

In these three constructors, **src** is a reference to the component by which the event was produced, and **type** specifies the type of the event. The argument **temporaryFlag** is set to **true,** if the focus event is temporary, else it is set to **false**.

 Example: Suppose that the focus event is in a text field. If the user moves the mouse to adjust a scroll bar, the focus is temporarily lost.

The other component involved in the focus change is called the opposite component, which is passed in **other**. Therefore, if a **FOCUS_GAINED** event occurs, **other** will refer to the component that lost focus. Conversely, if a **FOCUS_LOST** event occurs, **other** will refer to the component that gains focus.

Methods Used in the FocusEvent Class

1. ***getOppositeComponent( ):*** This method can be used to determine the other/opposite component. The syntax of this method is:

    Component getOppositeComponent( )

2. ***isTemporary( ):*** This method is used to specify whether the focus change is temporary or not. The syntax of this method is:

    Boolean isTemporary( )

    This method returns **true,** if the change is temporary, else it returns **false**.

14.3.6 The InputEvent Class

The abstract class **InputEvent** is a subclass of a **ComponentEvent** class and is the superclass for component input events. Its superclasses are **KeyEvent** and **MouseEvent**.

InputEvent defines several integer constants that represent modifiers, such as the control key being pressed and that might be associated with the event. The InputEvent class defines eight values to represent the modifiers as shown in table 14.5.

Table 14.5: Values to Represent Modifiers	
ALT_MASK	BUTTON2_MASK
ALT_GRAPH_MASK	BUTTON3_MASK
CTRL_MASK	BUTTON1_MASK
META_MASK	SHIFT_MASK

However, because of possible conflicts between the modifiers used by the keyboard events and mouse events and other issues, few extended modifier values were added. These extended values are shown in table 14.6.

Table 14.6: Extended Values to Represent Modifiers	
ALT_DOWN_MASK	BUTTON2_DOWN_MASK
ALT_GRAPH_DOWN_MASK	BUTTON3_DOWN_MASK
CTRL_DOWN_MASK	BUTTON1_DOWN_MASK
META_DOWN_MASK	SHIFT_DOWN_MASK

To test if a modifier was pressed at the time an event is generated, **isAltDown( )**, **isAltGraphDown( )**, **isControlDown( )**, **isMetaDown( )** and **isShiftDown( )** methods are used. The forms of the methods are:

Boolean isAltDown( )

Boolean isAltGraphDown( )

Boolean isControlDown( )

Boolean isMetaDown( )

Boolean isShiftDown( )

A value containing all the original modifier flags can be obtained by calling the **getModifiers( )** method. Syntax of this method:

int getModifiers( )

The extended modifiers can be obtained by calling **getModifiersEx( )**, which is shown as:

int getModifiersEx( )

14.3.7 The ItemEvent Class

An **ItemEvent** class is created when a list item or a check box is clicked or a checkable menu item is selected or deselected even once. Item events are of two types that are identified by the integer constants as shown in table 14.7.

Table 14.7: Integer Constants for ItemEvent Class	
DESELECTED	The user deselected an item
SELECTED	The user selected an item.

In addition, **ItemEvent** identifies an integer constant **ITEM_STATE_CHANGED**, that indicates a change of state.

Constructor of an ItemEvent Class

ItemEvent(ItemSelectable *src*, int *type*, Object *entry*, int *state*)

In this constructor, **src** is a reference to the component that generates the event, **type** specifies the type of the event. The specific item that created the item event is passed in **entry** and **state** specifies the current state of that item.

Methods Used in the ItemEvent Class

1. *getItem( ):* This method can be used to obtain a reference to the item that generated an event. Its syntax is:

    Object getItem( )

2. *getItemSelectable( ):* This method is used to get a reference to the **ItemSelectable** object that produced an event. Its syntax is:

    ItemSelectable getItemSelectable( )

 Example: Some of the examples of user interface essentials that implement the ItemSelectable interface are lists and choices.

3. *getStateChange( ):* This method is used to get the state change that is, SELECTED or DESELECTED for the event. Its syntax is:

    int getStateChange( )

14.3.8 The KeyEvent Class

A **KeyEvent** is generated when a keyboard input occurs. There are three types of key events that are identified by these integer constants

1. KEY_PRESSED
2. KEY_RELEASED
3. KEY_TYPED

When any key is pressed or released, the first two events are produced. The third event occurs when a character is typed.

Notes
It is not necessary that all keypresses produce characters. For example, pressing SHIFT does not generate a character.

There are many other integer constants that are defined by **KeyEvent** class. Table 14.8 shows some of them.

Table 14.8: ASCII Equivalents of Numbers and Letters

VK_ALT	VK_CANCEL
VK_DOWN	VK_LEFT
VK_RIGHT	VK_ENTER
VK_PAGE_DOWN	VK_SHIFT
VK_CONTROL	VK_ESCAPE
VK_PAGE_UP	VK_UP

Example: VK_0 through VK_9 and VK_A through VK_Z define the ASCII equivalents of the numbers and letters.

The VK constants indicate **virtual key codes** and are free from any modifiers, such as control, shift, or alt.

KeyEvent is a subclass of the **InputEvent** class. Its syntax is:

KeyEvent(Component *src*, int *type*, long *when*, int *modifiers*, int *code*, char *ch*)

In this syntax, **src** is a reference to the component that generates the event. The type of the event is specified by the **type**. The system time at which the key was pressed is passed in **when**. The **modifiers** argument indicates which modifiers were pressed when the key event occurred. In code, the **virtual key codes**, such as VK_UP, VK_A and so on are considered. The character equivalent is passed in **ch**. If
there is no existing valid character, then **ch** contains CHAR_UNDEFINED. The code contains VK_UNDEFINED in case of KEY_TYPED events.

Methods Used in the KeyEvent Class

1. *getKeyChar( ):* This method is used to get the character that was entered. Its syntax is:

   char getKeyChar( )

2. *getKeyCode( ):* This method is used to get the key code. Its syntax is:

   int getKeyCode( )

14.3.9   The MouseEvent Class

The MouseEvent class comprises eight types of mouse events. This class specifies integer constants that can be used to identify the event types shown in table 14.9.

Table 14.9: Integer Constants to Define the MouseEvent Class	
MOUSE_CLICKED	The user clicked the mouse.
MOUSE_DRAGGED	The user dragged the mouse.
MOUSE_ENTERED	The mouse entered a component.
MOUSE_EXITED	The mouse exited from a component.
MOUSE_MOVED	The mouse moved.
MOUSE_PRESSED	The mouse was pressed.
MOUSE_RELEASED	The mouse was released.
MOUSE_WHEEL	The mouse wheel was moved.

MouseEvent class is a child class of the **InputEvent** class. Its general form is as follows:

MouseEvent(Component *src*, int *type*, long *when*, int *modifiers*, int *x*, int *y*, int *clicks*, Boolean *triggersPopup*)

Here, **src** is a reference to the component that generates the event; **type** specifies the type of the event. The system at which the mouse event occurred is passed in **when**. The **modifiers** argument specifies which modifiers were passed when a mouse event occurred. The **x** and **y** are the two coordinates for the movement of mouse. The **clicks** specify the number of times a mouse is clicked and the **triggersPoPup** flag specifies whether this event leads to the occurrence of a pop-up menu, which appears on this platform.

Methods Used in the MouseEvent Class

1. *getX( ):* This method is used to get the **X** coordinate of the mouse within the component when the event occurred. Its syntax is:

    int getX( )

2. *getY( ):* This method is used to get the **Y** coordinate of the mouse within the component when the event occurred. Its syntax is:

    int getY( )

3. *getPoint( ):* The **getPoint( )** method can be used to get the coordinates of the mouse. Its syntax is:

    Point getPoint( )

    In the **getPoint( )** method, a Point object is returned, that contains the **X** and **Y** coordinates in its integer members, **x** and **y**.

4. *translatePoint( ):* This method is used to modify the event location. Its syntax is:

    void translatePoint(int x, int y)

    Here, the arguments **x** and **y** are the coordinates that are added to the event.

5. *getClickCount( ):* This method is used to get the number of mouse clicks for this event. Its syntax is:

   int getClickCount( )

6. *isPopupTriggers( ):* This method is used to check whether this event causes a pop-up menu to appear on this platform or not. Its syntax is:

   Boolean isPopupTrigger( )

7. *getButton( ):* This method is used to get a value that represents the button causing the event. Its syntax is:

   int getButton( )

The MouseEvent class defines the following constants:

1. NOBUTTON

2. BUTTON1

3. BUTTON2

4. BUTTON3

The NOBUTTON value shows that no button was pressed or released.

14.3.10 The MouseWheelEvent Class

The **MouseWheelEvent** class encloses mouse wheel events and is a child class of the **MouseEvent** class. Not all mice have wheels. If a mouse has a wheel, its (wheel's) location will be between the left and right buttons. Scrolling is done with the help of mouse wheels. The **MouseWheelEvent** class defines two integer constants, which are shown in the table 14.10.

Table 14.10: Integer Constants for MouseWheelEvent	
WHEEL_BLOCK_SCROLL	A page-up or page-down scroll event occurred.
WHEEL_UNIT_SCROLL	A line-up or line-down scroll event occurred.

Constructor of MouseWheelEvent Class

MouseWheelEvent (Component *src*, int *type*, long *when*, int *modifiers*, int *x*, int *y*, int *clicks*, Boolean *triggersPopup*, int *scrollHow*, int *amount*, int *count*)

In this constructor, **src** is a reference to the object that generates the event; **type** specifies the type of the event. The system at which the mouse event occurred is passed in **when**. The **modifiers** argument specifies which modifiers were passed when the event occurred. The **x** and **y** are the coordinates of the mouse that are passed. The **clicks** specify the number of times a mouse is clicked and the **triggersPoPup** flag specifies whether this event leads to the occurrence of a pop-up menu, which appears on this platform. The **scrollHow** values have to be either WHEEL_UNIT_SCROLL or WHEEL_BLOCK_SCROLL. The **amount** is a parameter, where the number of units to scroll is passed. The **count** parameter specifies the number of rotational units that the wheel moved.

Methods Used in the MouseWheelEvent Class

1. *getWheelRotation( ):* This method is used to get the number of rotational units. Its syntax is:

   int getWheelRotation( )

   A positive return value specifies the counterclockwise movement of the wheel, whereas a negative return value specifies the clockwise movement of the wheel.

2. *getScrollType( ):* This method is used to get the scroll type. Its syntax is:

    int getScrollType( )

    This method can either return WHEEL_UNIT_SCROLL or WHEEL_BLOCK_SCROLL.

3. *getScrollAmount( ):* This method is used to get the number of units to scroll. Its syntax is:

    int getScrollAmount( )

14.3.11 The TextEvent Class

Every instance of the **TextEvent** class specifies text events. When a user or a program enters a character, these text events are created by text fields and text areas. **TextEvent** class identifies the integer constant TEXT_VALUE_CHANGED.

Constructor of the TextEvent Class

TextEvent (Object *src*, int *type*)

In this constructor, **src** is a reference to the object that generates this event. The type of the event is specified in the parameter **type**.

The **TextEvent** object does not consist of the characters present in the text component that created the event. The object that executes the **TextListener** interface gets this **TextEvent** when the event arises. The details of handling individual mouse movements and key strokes are not done by the listener. But the listener can process a semantic event like **text changed**.

14.3.12 The WindowEvent Class

The WindowEvent class comprises ten types of window events. The **WindowEvent** class specifies integer constants that can be used for the identification of these events/this class. These constants are shown in the table 14.11.

Table 14.11 : Constants and their Meanings

WINDOW_ACTIVATED	The window was activated.
WINDOW_CLOSED	The window has been closed.
WINDOW_CLOSING	The window is requested (by the user) to be closed.
WINDOW_DEACTIVATED	The window was deactivated.
WINDOW_DEICONIFIED	The window was deiconified.
WINDOW_GAINED_FOCUS	The window gained input focus.
WINDOW_ICONIFIED	The window was iconified.
WINDOW_LOST_FOCUS	The window lost input focus.
WINDOW_OPENED	The window was opened.
WINDOW_STATE_CHANGED	The window's state was modified.

WindowEvent class is a child class of the **ComponentEvent** class. Its form is as follows:

WindowEvent(Window *src*, int *type*)

Here, **src** is a reference to the object that generates this event, and the **type** specifies the type of the event.

Constructors of the WindowEvent Class

1. WindowEvent (Window *src*, int *type*, Window *other*)

2. WindowEvent (Window *src*, int *type*, int *fromState*, int *toState*)

3. WindowEvent(Window *src*, int *type*, Window *other*, int fromState, int *toState*)

In these constructors, **other** specifies the opposite window, when a focus or activation event occurs. The **fromState** indicates the window's prior state, and **toState** indicates the new state that the window will have when a window state change takes place.

Methods Used in the WindowEvent Class

getWindow( ): This method is one of the frequently used methods in this class, which is used to get the Window object that generated the event. Its syntax is:

Window getWindow( )

WindowEvent also defines methods that return the opposite window (when a focus or activation event has occurred), the previous window state, and the current window state. These methods are:

1. Window getOppositeWindow( )

2. int getOldState( )

3. int getNewState( )

Write a simple program implementing the above mentioned event classes.

Task

14.4 Summary

*   Event models are an effective way of event handling in Java.

*   The event delegation model is considered as a modern approach to handle events.

*   In the event delegation model, an event refers to an object specifying a source's state change, and the source is an object generating that event.

*   An event listener is defined as an object that receives notification of the occurrence of an event..

*   An event is generated when the user interacts with a GUI application.

*   The subclasses of the AWT event can be categorized into two groups - semantic and low-level events.

*   Semantic events are defined as the events which directly correspond to high level user interactions with a GUI component. These events are ActionEvent, AdjustmentEvent, ItemEvent and TextEvent.

*   Every high-level user event may lead to multiple low-level events. Multiple low-level events are ComponentEvent, ContainerEvent, FocusEvent, InputEvent, KeyEvent, MouseEvent, MouseWheelEvent, PaintEvent and WindowEvent.

*   Event Listeners are objects and are used to handle a particular task of a component.

- Event delegation model can be used by following two simple steps. Firstly, execute the appropriate interface in the listener, and secondly, implement code to register and unregister the listener.

14.5 Keywords

ContainerEvent Class: A low level event that indicates that a container's content has changed after adding or removing a component.

FocusEvent Class: A low level event that indicates that a component has gained or lost the input focus.

KeyEvent Class: An event that indicates that a keystroke has occurred in a component.

MouseEvent Class: An event that indicates that a mouse action has occurred in a component.

Timestamp: The time of an event as recorded by a computer.

WindowEvent Class: A low level event which indicates that the status of a window has changed.

14.6 Self Assessment

1. State whether the following statements are true or false:

    (a) Event handling is an important aspect that relates to applets in Java.

    (b) At the root of the Java event hierarchy is EventObject, which is in java.util.

    (c) MouseWheelEvent is a Semantic event.

    (d) A ComponentEvent is generated when the size, position or visibility of a container is modified.

    (e) The advantage of the delegation event model is that the application logic that processes events is cleanly separated from the user interface logic that generates the events.

    (f) An Event listener is defined as an object that sends notification of the occurrence of an event.

    (g) The Listener implements the interface which contains event-handling code for a

    particular
    component.

2. Fill in the blanks:

    (a) In the event delegation model, the _____ generates an event and sends it to one or more listeners.

    (b) The method that registers a mouse motion listener is called _____.

    (c) The _____ class encapsulates a mouse wheel event.

    (d) The listeners are registered by an event source and are done by calling the _____ method.

    (e) A _____ is generated by a scroll bar.

3. Select a suitable choice in every question.

    (a) In the delegation event model, in order to receive an event notification the listeners must register with a source

        (i) source

        (ii) listener

        (iii) component

        (iv) event

(b)   Which one of the event is a semantic event?

  (i) FocusEvent

  (ii) ItemEvent

  (iii) PaintEvent

  (iv) ContainerEvent

(c)   Which one of the following is an integer constant for MouseWheelEvent?

  (i) WHEEL_BLOCK_SCROLL

  (ii) MOUSE_WHEEL

  (iii) WHEEL_BUTTON_SCROLL

  (iv) MOUSE_WHEELCLICKED

(d)   void mouseExited(MouseEvent me) is a constructor of which Interface?

  (i) The MouseMotionListener Interface

  (ii) The MouseWheelListener Interface

  (iii) The MouseListener Interface

  (iv) The MouseWheelEvent Class

(e)   The _____ method can be used to obtain a reference to the item that generated an event.

  (i) itemPoint( )

  (ii) itemEvent( )

  (iii) getStateChange( )

  (iv) getItem( )

14.7   Review Questions

1.   "In Java, event models are used for handling events". Explain.

2.   "The delegation event model is considered as a modern approach to handle events". Discuss.

3.   "An event source is an object that generates an event". Comment.

4.   "The subclasses of AWT events can be categorized into two groups". Elaborate.

5.   "A ComponentEvent is generated, when the size, position or visibility of a component is changed". Explain.

6.   "FocusEvent is a subclass of ComponentEvent and has three constructors". Discuss.

7.   "There are eight types of mouse events". Elaborate.

8.   "There are ten types of window events". Discuss.

9.   "Event Listeners are objects used to handle a particular task of a component". Discuss.

10.   "The interface specifies four methods that are invoked when a component is resized, moved, shown or hidden". Justify.

11.   "The mousePressed( ) and mouseReleased( ) methods are invoked when the mouse is pressed and released, respectively". Comment.

12.   "The WindowListener Interface specifies seven methods". Discuss.

Answers: Self Assessment

a.   (a) True   (b) True   (c) False

     (d) False   (e) True   (f) False   (g) True

b.   (a) Source   (b) addMouseMotionListener( )   (c) MouseWheelEvent

     (d) addXListener   (e) AdjustmentEvent.

c.   (a) Source   (b) ItemEvent   (c) WHEEL_BLOCK_SCROLL

     (d) The MouseListener Interface   (e) getItem( )

14.8 Further Readings

Books

Schildt. Herbert, The Complete Reference Java, Tata McGraw-Hill

Balagurusamy E. Programming with Java_A Primer 3e. New Delhi

Online link

http://notes.corewebprogramming.com/instructor/Java-Events.pdf

http://download.oracle.com/javase/1.4.2/docs/api/java/awt/event/package-tree.html

www.ingramcontent.com/pod-product-compliance
Lightning Source LLC
Chambersburg PA
CBHW031828170526
45157CB00001B/224